自动控制英语
ENGLISH FOR AUTOMATIC CONTROL

主　编　王　昕　张　静　牛存超

北京理工大学出版社
BEIJING INSTITUTE OF TECHNOLOGY PRESS

内 容 简 介

本教材以学生为中心，以课程思政为引领，以成果输出为导向，以任务完成为驱动，将语言知识落实到真实的职场任务中。

本教材内容围绕自动控制岗位核心业务知识与流程，设置八个主题，即电气工程、计算机辅助设计和计算机辅助制造、自动化、工作场所安全、自动驾驶汽车、高速列车、机器人和智慧城市。本教材以职场环境中的典型工作任务为主线，以杨康这一职场人物为主角，以杨康所遇到的专业问题为切入点，创设交际情境，提升学生在职场环境中英语运用能力和沟通能力。

本教材充分利用先进的信息化教学手段，为学生提供丰富便捷的数字资源和学习路径，包括授课视频、微课、音频、课件等。通过本课程学习，学生将掌握一定的自动控制行业英语知识，能够开展高效的职场交际活动，提升职业素养和工作适应性。本教材适用于高等职业院校或初等职业院校自动控制相关专业的学生或从事该专业的在职人员。

版权专有　侵权必究

图书在版编目(CIP)数据

自动控制英语 / 王昕，张静，牛存超主编. -- 北京：北京理工大学出版社，2024.1
ISBN 978 - 7 - 5763 - 3539 - 2

Ⅰ. ①自… Ⅱ. ①王… ②张… ③牛… Ⅲ. ①自动控制 - 英语 - 高等职业教育 - 教材 Ⅳ. ①TP273

中国国家版本馆 CIP 数据核字(2024)第 042923 号

责任编辑 / 王梦春	文案编辑 / 辛丽莉
责任校对 / 周瑞红	责任印制 / 施胜娟

出版发行 / 北京理工大学出版社有限责任公司
社　　址 / 北京市丰台区四合庄路 6 号
邮　　编 / 100070
电　　话 / (010) 68914026（教材售后服务热线）
　　　　　 (010) 68944437（课件资源服务热线）
网　　址 / http://www.bitpress.com.cn
版 印 次 / 2024 年 1 月第 1 版第 1 次印刷
印　　刷 / 唐山富达印务有限公司
开　　本 / 787 mm × 1092 mm　1/16
印　　张 / 14
字　　数 / 298 千字
定　　价 / 70.00 元

图书出现印装质量问题，请拨打售后服务热线，负责调换

前 言

为贯彻落实《中共中央关于认真学习宣传贯彻党的二十大精神的决定》《习近平新时代中国特色社会主义思想进课程教材指南》《职业院校教材管理办法》等文件精神，本教材编写团队认真执行思政内容进教材、进课堂、进头脑的要求，尊重教育规律，遵循学科特点，对教材内容进行了编写。

一、编写宗旨及特色

（1）提升教材铸魂育人功能，培育、践行社会主义核心价值观，教育引导学生树立共产主义远大理想和中国特色社会主义共同理想。同时，弘扬优秀中华传统文化。

（2）注重科学思维方法的训练和科学伦理教育，培养学生探索未知、追求真理、勇攀科学高峰的责任感和使命感；培养学生精益求精的大国工匠精神，培养学生科技报国的家国情怀和使命担当。

（3）教育引导学生深刻理解并自觉实践各行业的职业精神、职业规范，增强职业责任感，培养遵纪守法、爱岗敬业、无私奉献、诚实守信、公道办事、开拓创新的职业品格和行为习惯。

（4）教材知识内容及时更新，体现产业发展的新技术、新工艺、新规范、新标准。加强教材数字化建设，丰富配套资源，形成可听、可视、可练、可互动的融媒体教材。

二、内容简介

本教材内容包含了从基础理论到具体应用的各类专业知识，共设置八个主题：电气工程、计算机辅助设计和计算机辅助制造、自动化、工作场所安全、自动驾驶汽车、高速列车、机器人和智慧城市。每个单元包括以下几个部分。

1. 导入

倡导听说领先。通过浏览与单元主题相关的图片及听简短的对话了解该单元的内容，以讨论的方式进入学习，调动学生的主观能动性。

2. 对话

通过相关专业真实语境，调整小组成员并反复练习情景对话；通过记忆训练和模仿练习，有效地促成行业语言板块的形成。

3．课文

语言真实规范，题材新颖，密切联系实际，没有艰难晦涩、专业难度过大的内容；篇幅短小实用，一般在 200 字左右，避免篇幅过长带来沉重感，从而消除学生的畏惧感。

4．翻译技巧

介绍常见的科技英语翻译方法与技巧，通过理论知识讲解和大量实例展示，提高学生的翻译能力。

5．写作

将常见的应用文文体与商务公文写作相结合，通过一系列的范例学习和写作实践来培养学生阅读和模拟套写常用应用文的能力。

6．文化点滴

拓展与主题相关的背景知识，增加学生对英语学习的兴趣及对本行业的热爱，开阔视野。

7．自评表格

每个单元后的自我评价表有助于学生提升自信、增强自我认知和自我控制能力，对学生发展有重要的意义与作用。

三、任务分工

本教材由辽宁机电职业技术学院的王昕、张静、牛存超编写。其中，第五单元、第六单元和第八单元（包括参考答案）及前言由王昕编写，第一单元、第二单元和第七单元（包括参考答案）由张静编写，第三单元、第四单元（包括参考答案）由牛存超编写。

本教材在编写过程中参考了大量的文献资料，同时也得到了各方面的支持和帮助。在此，我们向所有的作者表示感谢！也感谢领导、同事及家人的默默的支持和不断的鼓励！

虽然编者已竭尽全力，但由于水平有限，教材中难免有疏漏和不妥之处，恳请广大读者批评指正。

编 者

目录 Contents

Unit One Electrical Engineering ········· 1
Section One Warming Up ········· 2
Section Two Dialogue: Am I Qualified to Be an Electrician? ········· 3
Section Three Passage: Electrical Engineers ········· 7
Section Four Translation Skills ········· 13
Section Five Writing: A Cover Letter ········· 16
Section Six Culture Tips ········· 19
Section Seven Self-evaluation ········· 20

Unit Two CAD & CAM ········· 21
Section One Warming Up ········· 22
Section Two Dialogue: Introduction to CAD and CAM ········· 23
Section Three Passage: Application of CAD Technology ········· 27
Section Four Translation Skills ········· 33
Section Five Writing: A Company Introduction Letter ········· 36
Section Six Culture Tips ········· 39
Section Seven Self-evaluation ········· 40

Unit Three Automation ········· 41
Section One Warming Up ········· 42
Section Two Dialogue: Visiting an Automated Factory ········· 43
Section Three Passage: Advantages and Disadvantages of Automation ········· 46
Section Four Translation Skills ········· 53
Section Five Writing: An Event Invitation Letter ········· 56
Section Six Culture Tips ········· 59
Section Seven Self-evaluation ········· 59

Unit Four　Workplace Safety ····· 61
　Section One　Warming Up ····· 62
　Section Two　Dialogue：Talking about Electrical Safety ····· 63
　Section Three　Passage：Safety Rules in the Workshop ····· 66
　Section Four　Translation Skills ····· 72
　Section Five　Writing：A Quotation Request Letter ····· 75
　Section Six　Culture Tips ····· 78
　Section Seven　Self-evaluation ····· 79

Unit Five　Autonomous Cars ····· 80
　Section One　Warming Up ····· 81
　Section Two　Dialogue：How Far Away Are We from Autonomous Cars? ····· 82
　Section Three　Passage：Google's Self-driving Vehicles ····· 85
　Section Four　Translation Skills ····· 92
　Section Five　Writing：A Shipment Request Letter ····· 95
　Section Six　Culture Tips ····· 98
　Section Seven　Self-evaluation ····· 99

Unit Six　High Speed Trains ····· 100
　Section One　Warming Up ····· 101
　Section Two　Dialogue：History of High Speed Trains ····· 102
　Section Three　Passage：High Speed Trains in China ····· 105
　Section Four　Translation Skills ····· 111
　Section Five　Writing：A Late Delivery Apology Letter ····· 114
　Section Six　Culture Tips ····· 117
　Section Seven　Self-evaluation ····· 118

Unit Seven　Robots ····· 119
　Section One　Warming Up ····· 120
　Section Two　Dialogue：Do You Like Robots to Work at Your Home? ····· 121
　Section Three　Passage：Industrial Robots ····· 124
　Section Four　Translation Skills ····· 130
　Section Five　Writing：A Business Thank You Letter ····· 133
　Section Six　Culture Tips ····· 135
　Section Seven　Self-evaluation ····· 136

Unit Eight　A Smart City ····· 137
　Section One　Warming Up ····· 138
　Section Two　Dialogue：What Does a Smart City Look Like? ····· 139
　Section Three　Passage：Smart Cities and Normal Cities—What Are the Differences? ····· 143
　Section Four　Translation Skills ····· 149

Section Five　Writing：A Job Promotion Congratulation Letter …………………… 151

Section Six　Culture Tips ………………………………………………………………… 154

Section Seven　Self-evaluation ………………………………………………………… 155

参 考 文 献 …………………………………………………………………………………… 156

Answer Keys …………………………………………………………………………………… 157

Unit One　Electrical Engineering ……………………………………………………… 157

Unit Two　CAD & CAM ………………………………………………………………… 161

Unit Three　Automation ………………………………………………………………… 164

Unit Four　Workplace Safety …………………………………………………………… 168

Unit Five　Autonomous Cars …………………………………………………………… 172

Unit Six　High Speed Trains …………………………………………………………… 176

Unit Seven　Robots ……………………………………………………………………… 180

Unit Eight　A Smart City ………………………………………………………………… 184

Listening Scripts ……………………………………………………………………………… 189

Unit One　Electrical Engineering ……………………………………………………… 189

Unit Two　CAD & CAM ………………………………………………………………… 192

Unit Three　Automation ………………………………………………………………… 195

Unit Four　Workplace Safety …………………………………………………………… 198

Unit Five　Autonomous Cars …………………………………………………………… 201

Unit Six　High Speed Trains …………………………………………………………… 204

Unit Seven　Robots ……………………………………………………………………… 207

Unit Eight　A Smart City ………………………………………………………………… 210

敬业乐群。
Work diligently and keep good company with others.

——《礼记》
—The Book of Rites

Unit One

Electrical Engineering

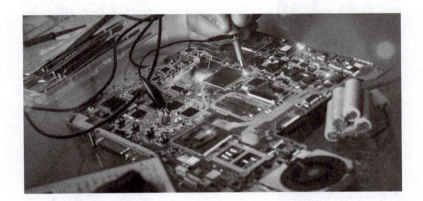

Focus

Section One	Warming Up
Section Two	Dialogue: Am I Qualified to Be an Electrician?
Section Three	Passage: Electrical Engineers
Section Four	Translation Skills
Section Five	Writing: A Cover Letter
Section Six	Culture Tips
Section Seven	Self-evaluation

Learning Objectives

Upon completion of the unit, students will be able to:

1. Have basic knowledge of electrical engineering.

2. Fully understand the essential qualifications for electricians and electrical engineers.

3. Keep in mind the steps that you can take to improve your electrical engineering skills.

Section One Warming Up

Group Work

Do you know the following famous persons and their contributions in electricity?

A. Inventor of lightning rod.
B. Ohm's Law.
C. Inventor of voltaic pile.
D. Father of alternating current.

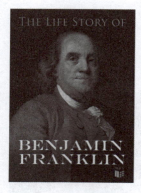

Listen to the dialogue and find out which person above they are talking about.

Warming Up 听力

Unit One Electrical Engineering

Section Two Dialogue: Am I Qualified to Be an Electrician?

Listen and role-play the following dialogue.

Yang Kang: Look at the job advertisement. The Electric Power Bureau is looking for a **professional** and experienced electrician. Do you think I am qualified for it?

Sally: **Absolutely**. You have completed the electrical training and an **apprenticeship**. And also, you've obtained the senior electrician **certificate**.

Yang Kang: It seems that more electricians are **currently** in demand and well-paid.

Sally: Because electricity is such a **vital** part of modern-day society. We need electricians in a **variety** of locations, from businesses, **residential properties**, and **commercial** projects, to **construction** zones. They are necessary parts of our society.

Yang Kang: **Installation**, fitting, rewiring, repairing, testing, and **maintaining** electrical systems and **components** are all tasks an electrician is responsible for. It can involve anything from fixing a light bulb to maintaining a solar panel. I wish I had a more solid understanding of electrical terms.

Sally: Since technology is **constantly** one step ahead and moving at a rapid pace, electricians are **perpetual** students. To stay in touch with the growth, and to understand how all the new processes work, they have to take training courses.

Yang Kang: Exactly! I couldn't agree more. Besides that, having good **observational** skills will be helpful to **identify** electrical problems, deal with various problems and avoid accidents.

Sally: Are you confident of getting the job now?

Yang Kang: Yes.

Words & Expressions

我有资格成为电工吗?

qualified	[ˈkwɒlɪfaɪd]	adj. 有资格的；能胜任的
electrician	[ɪˌlekˈtrɪʃn]	n. 电工；电气技师
professional	[prəˈfeʃnl]	adj. 职业的；专业的 n. 专业人员
absolutely	[ˈæbsəluːtli]	adv. 完全地；绝对地

		int. （表示赞同）一点没错
apprenticeship	[əˈprentɪʃɪp]	n. 学徒期；学徒身份；学徒资格
certificate	[səˈtɪfɪkeɪt]	n. 证明；合格证书；文凭
currently	[ˈkʌrəntli]	adv. 现在；目前
vital	[ˈvaɪtl]	adj. 至关重要的；有活力的
variety	[vəˈraɪəti]	n. 多样；种类；多样化
residential	[ˌrezɪˈdenʃl]	adj. 住宅的；居住的
property	[ˈprɒpəti]	n. 财产；所有物；地产
commercial	[kəˈmɜːʃl]	adj. 商业的 n. 商业广告
construction	[kənˈstrʌkʃn]	n. 建设；结构；建筑物
installation	[ˌɪnstəˈleɪʃn]	n. 安装；装置
maintain	[meɪnˈteɪn]	vt. 维持；维修；保养
component	[kəmˈpəʊnənt]	n. 零部件；元件；组成部分
constantly	[ˈkɒnstəntli]	adv. 不断地；经常地
perpetual	[pəˈpetʃuəl]	adj. 永久的；一再往复的
observational	[ˌɒbzəˈveɪʃənl]	adj. 观察的；观测的
identify	[aɪˈdentɪfaɪ]	vt. 识别；显示；说明身份

job advertisement	招聘广告
solar panel	太阳能电池板
electrical term	电气术语

 Find Information

Task Ⅰ. Read the dialogue carefully and then answer the following questions.

1. Why is Yang Kang attracted by the job advertisement?
_____

2. What is required to be an electrician?
_____

3. What's the present situation of electricians?
_____

4. Where do electricians usually work?
_____

5. Is Yang Kang qualified to be an electrician?
_____

Unit One Electrical Engineering

Task Ⅱ. Read the dialogue carefully and then decide whether the following statements are true (T) or false (F).

(   ) 1. Yang Kang wonders whether he is qualified to be an electrician.
(   ) 2. Yang Kang is very interested in the job advertisement.
(   ) 3. Electricity is vital in the modern society.
(   ) 4. Electricians are not perpetual students.
(   ) 5. Electricians need to identify and deal with various electrical problems.

Cheer up Your Ears

Listen and write down what you've heard. Then read and recite till you can use them fluently.

1. The Electric Power Bureau is looking for a _____ electrician.
2. Yang Kang has completed the electrical training and an _____ .
3. He obtained the senior electrician _____ .
4. Electricity is such a _____ part of modern-day society.
5. We need electricians in a _____ of locations.
6. An electrician is responsible for installation, fitting, _____, repairing, testing, etc.
7. It can involve anything from fixing a light bulb to _____ a solar panel.
8. Technology is _____ one step ahead and moving at a rapid pace.
9. Having good _____ skills will be helpful to avoid accidents.
10. Are you _____ of getting the job now?

DIALOGU ECHEER
UP EARS

Words Building

Task Ⅰ. Choose the best answer from the four choices A, B, C and D.

(   ) 1. The hopes, goals, fears and desires _____ widely between men and women, between the rich and the poor.
  A. alter   B. shift   C. transfer   D. vary

(   ) 2. His professional career _____ 16 years.
  A. spanned   B. spaned   C. is spanned   D. spannes

(   ) 3. Now I am not sure if I will _____ to graduate next semester.
  A. qualifier   B. be qualified C. qualify   D. be qualifying

(   ) 4. The price of _____ has risen enormously.
  A. property   B. prosperity   C. prospect   D. proposal

(   ) 5. These carpets are more _____ demand by the salaried class.
  A. on   B. for   C. out of   D. in

Task Ⅱ. Fill in each blank with the proper form of the word given.

1. She was in the second year of her _____ as a carpenter. (apprentice)

2. We need an _____ to fix our light switch. (electricity)

3. Within ten weeks of the introduction, 34 million people would have been reached by our television _____. (commercial)

4. The _____ situation is very frustrating for us. (currently)

5. Careful _____ can extend the life of your car. (maintain)

Task Ⅲ. Match the following English terms with the equivalent Chinese items.

A. electrician certificate   1. (　　) 意外保险

B. intellectual property   2. (　　) 学徒培训

C. accident insurance   3. (　　) 电路元件

D. professional staff   4. (　　) 身份证

E. apprenticeship training   5. (　　) 施工程序

F. installation instructions   6. (　　) 电工证

G. circuit components   7. (　　) 万年历

H. perpetual calendar   8. (　　) 专业人士

I. construction program   9. (　　) 知识产权

J. identification card   10. (　　) 安装说明

Table Talk

Pair Work

Role-play a conversation about being an electrician with your partner.

Situation

Imagine you want to be a new electrician in a company. You are interviewed by the interviewer. Now you are talking with him.

Unit One Electrical Engineering

Section Three Passage: Electrical Engineers

Pre-reading Questions

1. What do you know about electrical engineering?
2. What qualities do you think are necessary for an electrical engineer?
3. Would you like to be an electrical engineer in the future?

电气工程师

In today's digital age, electricity truly keeps the world running, from basics of maintaining our homes to the more **complex** systems of traffic lights, **transportation** and technology that keep our cities running.

Electrical engineers are the designers that create these systems and keep them running **smoothly**, working on everything from the nation's power grid to the **microchips** inside our cell phones and smart watches. Electrical engineers are involved in designing, developing, testing and directing the manufacture of electrical equipment, such as **radar** and **navigation** systems, electric motors, **generators**, or communication systems. They are also responsible for designing **aviation** and **automotive** systems.

As an electrical engineer, you have the opportunity to **impact** the way that the world works. Here are some steps you can take to improve your electrical engineering skills. First of all, focus on schoolwork. Be familiar with **expertise**. Engineering courses also often teach you the soft skills you'll need, such as communication and cooperation. **Furthermore**, it can be helpful to attend training courses as much as possible to **enhance** your skills. **Additionally**, you can also ask your **colleagues** and **clients** for **feedback** on your work to determine areas that need improvement. Feedback can help you identify those areas and decide how you'll improve in the future. Finally, pursue **certifications** available for electrical engineers.

Words & Expressions

complex	[kəmˈpleks]	adj. 复杂的；合成的 n. 综合体；复合体
transportation	[ˌtrænspɔːˈteɪʃn]	n. 运输；运输工具；运输系统
smoothly	[ˈsmuːðli]	adv. 平稳地；流畅地
microchip	[ˈmaɪkrəʊtʃɪp]	n. [计] 微芯片
radar	[ˈreɪdɑː]	n. 雷达

navigation	[ˌnævɪˈgeɪʃn]	n. 导航；航行；航海
generator	[ˈdʒenəreɪtə]	n. 发电机
aviation	[ˌeɪviˈeɪʃn]	n. 航空；飞机制造业
automotive	[ˌɔːtəˈməʊtɪv]	adj. 汽车的；机动的
impact	[ˈɪmpækt]	n. 影响；冲击力 vt./vi. 撞击；产生影响
expertise	[ˌekspɜːˈtiːz]	n. 专门知识；专门技术
furthermore	[ˌfɜːðəˈmɔː]	adv. 而且；此外
enhance	[ɪnˈhæns]	vt. 增强；提高
additionally	[əˈdɪʃənəli]	adv. 另外
colleague	[ˈkɒliːg]	n. 同事
client	[ˈklaɪənt]	n. 委托人；客户
feedback	[ˈfiːdbæk]	n. 反馈；反馈意见
certification	[ˌsɜːtɪfɪˈkeɪʃn]	n. 证书；证明；鉴定

digital age	数字时代
power grid	电力网
electric motor	电机
communication system	通信系统

 Find Information

Task Ⅰ. Read the passage carefully and then answer the following questions.

1. Why is electricity playing a significant role in our society?

2. What is the main purpose of electrical engineers?

3. Which areas are electrical engineers engaged in?

4. What skills do electrical engineers need?

5. Do you think electrical engineers contribute a lot to the world?

Task Ⅱ. Read the passage carefully and then decide whether the following statements are true（T）or false（F）.

(   ) 1. Some electrical engineers design and maintain power plants, transmission lines and home and factory electrical installations.

(   ) 2. Some electrical engineers contribute to designing aviation and automotive systems.

(   ) 3. Electrical engineers help to drive the whole world all along.

(   ) 4. Having expertise and skills in electrical engineering will give you more career options.

(   ) 5. Taking training courses and obtaining certifications is an essential part of career growth.

Cheer up Your Ears

Listen and write down what you've heard. Then read and recite till you can use them fluently.

1. Electricity truly keeps the world running, from basics of _____ our homes to the more complex systems that keep our cities running.

2. Electrical engineers create these systems and keep them running _____.

3. They work on everything from the nation's power grid to the microchips inside our cell phones and _____.

4. Electrical engineers are involved in designing, developing, testing and directing the _____ of electrical equipment.

5. They are also _____ for designing aviation and automotive systems.

6. Electrical engineers have the opportunity to _____ the way that the world works.

7. Focus on schoolwork. Be familiar with _____.

8. Engineering courses also often teach you the soft skills you'll need, such as _____ and cooperation.

9. It can be helpful to attend training courses to _____ your skills.

10. Pursue _____ available for electrical engineers.

Words Building

PASSAGE CHEER UP EARS

Task Ⅰ. Translate the following phrases into English or Chinese.

1. 电气设备 _____
2. 交通灯 _____
3. 手机和智能手表 _____
4. 电机 _____
5. 通信系统 _____
6. power grid _____
7. radar and navigation _____
8. digital age _____
9. training courses _____

10. automotive systems _____

Task Ⅱ. Choose the best answer from the four choices A, B, C and D.

(   ) 1. I knew very well that the problem was _____ than he supposed.
A. more complex　　　　　　　　　　B. complexer
C. much complex　　　　　　　　　　D. more complexly

(   ) 2. They recommend that a new committee _____ .
A. creates　　　B. created　　　C. be created　　　D. create

(   ) 3. We should take pains to make our products _____ to wide market.
A. smooth　　　B. impact　　　C. transport　　　D. available

(   ) 4. _____ these arrangements, extra ambulances will be on duty until midnight.
A. In addition to　　　　　　　　　　B. Additionally
C. In addition　　　　　　　　　　　D. In addition with

(   ) 5. There was an _____ of 42 at the meeting.
A. attendence　　　　　　　　　　　B. attending
C. attendance　　　　　　　　　　　D. attendant

Task Ⅲ. Fill in each blank with the proper form of the word given.

1. Their medical problems are _____ physical in origin. (basic)
2. It would be better to _____ the goods by rail rather than by road. (transportation)
3. People have to take _____ for themselves and their lives. (responsible)
4. I suspect that he is more or less _____ in the affair. (involve)
5. In any event, you should choose a reputable _____ . (manufacture)

Task Ⅳ. Complete the sentences below with the correct form of the words and phrases in the box.

| client | generator | automotive | power |
| certifications | feedback | expertise | colleagues |

1. It can provide students with instant _____ , including reports about their strengths and weaknesses.
2. The affair was settled to the complete satisfaction of the _____ .
3. We were friends and _____ for more than 20 years.
4. You are an expert on something and you should share your _____ with educators and their students.
5. A _____ failure plunged everything into darkness.
6. How much do _____ help when someone is looking for a job?
7. Each home will have a solar _____ to provide power for lighting and heating.
8. Chinese car brands are slowly making an impact on the Australian _____ market.

Unit One Electrical Engineering

Task V. Translate the following sentences into Chinese.

1. In today's digital age, power systems keep our cities running.

2. Electrical engineers are involved in designing, developing, testing and directing the manufacture of electrical equipment.

3. Electrical engineers are supposed to learn expertise well.

4. Furthermore, it can be helpful to attend training courses as much as possible to enhance skills.

5. You can also ask your colleagues and clients for feedback on your work to determine areas that need improvement.

 Listening

Task Ⅰ. Listen to five short dialogues and choose the best answer.

1. A. A football game.   B. A rock concert.
   C. A new film.   D. A talk show.
2. A. A computer programmer.   B. An office secretary.
   C. An assistant manager.   D. A chief engineer.
3. A. In a bookstore.   B. In a library.
   C. In a supermarket.   D. In a restaurant.
4. A. Cash a check.   B. Open an account.
   C. Change some money.   D. Make a deposit.
5. A. He is selling a product.   B. He is looking for a job.
   C. He is making a welcome speech. D. He is interviewing a job applicant.

A级听力单选

A级听力填空

Task Ⅱ. Listen to the passage and fill in the blanks with the missing words or phrases.

How great it is to see so many of you come and join us in celebrating the 15th anniversary of our travel magazine. From the bottom of 1. _____, we thank you for being here. A little more than fifteen years ago, we were sitting at our regular jobs, 2. _____ how we saw our future, when we came up with the idea of joining our two hobbies, traveling and writing. We never imagined that our tiny dream would 3. _____ so soon. There were many special people who joined us and made it 4. _____ to create the name that we have today. To all those people and those who joined us in our journey, I should say thank you again. 5. _____ we would never have been here.

 **Extensive Reading**

Directions: *After reading the passage, you are required to choose the best answer from the four choices for each statement.*

Electrical System Design

Electrical design deals with power transition, distribution of the power, lighting system, telecommunication, fire alarm system, public addressable system and closed-circuit television.

To become good at designing electrical systems, here are a few things you must consider.

1. Have a basic knowledge on electronics.

Before you can start designing an electric system, you need to understand the electrical components, their interphases, and fundamentals.

2. Emphasize on connectivity.

We are living in an interconnected world. Most devices have several antennas. Along with high-frequency signals from WiFi, they might include transducers that might contribute to potential interphase issues. Hence, while designing an electrical system, you must include optimization of the 5G network, internet of things (IOT), and dedicated campus network.

3. Mechanical requirements.

Electrical designs are meant to withstand a certain level of structural and operational requirements. Some problems might challenge the overall design of the electrical system.

4. Thermal condition of the system.

When an electrical system performs, it tends to get heated up. Your design must negate the thermal conditioning of the system. You must conduct thermal flow analysis.

5. Circuit simulation.

Circuit simulation plays a strategic role in the schematic phase. During the circuit simulation process, you will be able to find the errors and resolve the issues before the system is launched.

6. Coding.

To develop a complex electrical system, a design engineer needs software for model-based development.

7. Mechatronics.

Today, every product needs to be smart in every way possible. Smart sensors and actuators are strategic components for smart product monitoring and control. In addition, these smart technologies help the users deliver the performance needed for the smart system.

1. Which three aspects of electrical design will help you with a solid start?

A. Electrical components. B. Interphases.

C. Fundamentals. D. A, B and C.

2. When designing an electrical system, you must include the following except _____.
   A. optimization of the 5G network B. dedicated campus network
   C. internet of things D. smart product monitoring and control
3. What plays a strategic role in the schematic phase? _____.
   A. Circuit simulation B. Mechatronics
   C. Coding D. Thermal condition of the system
4. To develop a complex electrical system, a design engineer needs _____.
   A. upgradaing software
   B. software for model-based development
   C. coding
   D. hardware for model-based development
5. What are strategic components for smart product monitoring and control?
   A. Smart sensors and actuators. B. Smart systems.
   C. Complex software. D. Circuits.

Section Four　Translation Skills

科技英语的翻译方法与技巧——照应译法

在进行科技英语翻译时,译文可以与原语序保持一致,按照字面意思进行翻译,使译文既符合原文的结构形式,也忠于原文的内容,这种从前往后的翻译方法,就是照应译法。

一、词语照应

科技英语中存在大量"术语"或"行话"。照应译法具有概念明确等优点,广泛应用于科技术语的翻译中,可以从音、形和义三个方面进行。

(一) 音译法

当英、汉两种语言某些词语差异很大,存在语义空缺,翻译无法直接从形式或语义方面进行加工转换时,音译则是主要翻译手段。音译要力求保持原文的读音,字、词尽可能保留英语的习惯。音译对象主要是人名和地名。如:

Fourier transform 傅里叶变换　　　Williams tube 威廉姆斯管
Brinell hardness 布氏硬度　　　　Vickers hardness 维氏硬度
Babbitt metal 巴氏合金　　　　　Helmholtz coil 亥姆霍兹线圈

另外,科技发展所带来的新产品、新概念、新理论等词语,也可以借助音译法翻译。如:

radar 雷达　　sonar 声呐　　Hertz 赫兹　　gene 基因

（二）象形译法

通过具体的形象，直接表达原意，如果在汉语中能找到对等词，就进行完全照应翻译。如：

Y-intersection 三岔路口　　　　　　X-type 交叉梁
T-square 丁字尺　　　　　　　　　　I-bam 工字梁
A-connection 三角接法　　　　　　　Y-connection 星形接法
V-connection V 形接法　　　　　　　V-shaped neck line V 形领口

对于科技英语中所涉及的商标名称、牌号、型号以及代表某种概念或具有特定意义的字母，也可选择象形译法，保留原英文字母不作翻译，只需要将普通名词翻译出即可。如：

V-neck sweater V 领（鸡心领）毛衣　　X-stream X 流

（三）直译法

把术语的各个组成部分的意义按顺序依次翻译出来。如：

firewall 防火墙　　　　　　　　　　handicraft 手工艺（品）
hardware 硬件　　　　　　　　　　　loudspeaker 扬声器
mooncraft 月球飞船　　　　　　　　 software 软件
artificial intelligence 人工智能　　　　artificial leather 人造革
large-scale industry 大型工业　　　　 superconductor 超导体
human-machine interface 人机界面　　artificial neural network 人工神经网络

有时，科技发展所产生的新的英语术语在汉语中找不到更合适的对应项，这时也可以按照字面意义直译，赋予新术语各组成部分在新语境中的新生意义，达到丰富汉语词汇的目的。如：

cyber café 网吧　　　　　　　　　　Internet surfing 网上冲浪
floppy disc 软盘　　　　　　　　　　information superhighway 信息高速公路

在音译、象形译和直译这三种照应译法中，直译法是科技术语翻译中使用最为普遍的一种。

二、句子照应

（一）概述

在英、汉句子结构中，主要句子成分顺序基本相同，常见的有以下四种。

1. 主语 + 谓语（不及物动词或系表结构），如：

The sewing machine won't run properly.
【译文】缝纫机不能正常运转了。
Clean energy is cost effective for governments and enterprises.
【译文】清洁能源对政府和企业都有利。

2. 主语+谓语+宾语，如：

For one thing, they simplify the design of a rocket engine.

【译文】首先，它们简化火箭发动机的设计。

3. 主语+谓语+宾语+宾语补足语，如：

A rotor from the engine makes the wheels of the car turn.

【译文】发动机的转子驱动车轮。

4. 主语+谓语+间接宾语+直接宾语，如：

National economy's hope of renewable energy gives wind power industry vast room for growth.

【译文】国民经济对再生能源的期许，给风电产业提供了广阔的发展空间。

（二）照应翻译

由于人类的思维模式具有一定的相似性，在语言表达时，也往往会遵循动作发生的先后顺序、因果、假设等逻辑关系，所以在科技英语翻译中，尽量采用照应译法，有助于信息的有效传达和理解。

1. 简单句照应翻译。

如果所要翻译的英语句子只是一个简单句，可以按照原文主谓顺序翻译，无须改变原文的顺序或结构。如：

Mechanized data-processing techniques have been successfully applied to index making.

【译文】机械化数据处理技术已经成功地应用于索引的编制。

How to select the cutting tool depends on the material to be machined.

【译文】怎样选择刀具取决于所要加工的材料。

2. 复合句照应翻译。

科技英语中有大量的通过"连接词"将两个或两个以上的子句组合而构成的复合句。这些复合句中常见的连接词有表示并列关系的 and、but、or 等；表示主从关系的 when、why、though 等；表示转折关系的 however、though、although 等；表示让步关系的 even if、even though、if 等；表示时间关系的 when、while、as、until、till、as soon as 等；表示假设关系的 as if、as though、once 等；表示条件关系的 if、in case、provided that 等；表示选择关系的 whether 等。

对于并列句，只需要抓住"连接词"这一连接点，先将连接词左右分句分别照应翻译，然后根据它们的逻辑关系组合即可。如：

Technical language and jargon are useful as professional shorthand, but they are barriers to successful communication with the public.

【译文】专业术语和行话有助于专业速记，但是会妨碍与公众的顺畅交流。

Renewable feedstocks are often made from agricultural products or are the wastes of other processes, and depleting feedstocks are made from fossil fuels.

【译文】可再生原料通常源自农作物或其他制作过程的废弃物，而消耗性原料来自石化燃料。

对于主从关系的复合句，照应翻译主要应用于下列两种情形。

第一，时间、原因、条件及让步等状语从句在前，主句在后的复合句。till 引导的从句可以跟在主句之后。如：

If these chemicals are necessary, use innocuous chemicals.

【译文】如果这些化学药品是必需的，就使用无害的化学药品。

Whatever the configuration of a chilled water system is, proper control is necessary to meet the overall system requirements.

【译文】不管冷却水系统配置如何，都必须进行合理的控制，从而满足整个系统的要求。

第二，只包含名词性从句的复合句。如：

What the consumer ultimately decides to purchase is influenced by other factors.

【译文】消费者最终决定购买何种商品受其他因素的影响。（主语从句，采用照译法。）

This is why this aspect is best left to the professional graphics designers.

【译文】这就是为什么最好将这方面内容留给专业平面设计师去做的原因。（why 引导表语从句，采用照译法。）

Section Five　Writing：A Cover Letter

求职信

一、写作要点 Key Points

Step 1：说明欲应聘的职务名称，以及在何处得知招聘信息。

Step 2：陈述过去的工作经验，以及自己具备哪些符合职务需求的专业特质。

Step 3：感谢对方考虑，并表明希望能有进一步面试的机会。

二、实际 E-mail 范例

| 写信▼ | 删除 | 回复▼ | 寄件者： | Ginny |

Dear Sir/Madam,

Introduction

Please find the attachment for my resume in application for the post advertised on Manpower Agency on May 15.

Body

As you can see in my resume, I have had extensive work experience in retail sales and service industries, where I gained skills and the ability to work with different types of customers and deal with complaints. In addition to these, I am also proficient in meeting sales targets. Also, I am very flexible and adapt well to different situations and new environments. I am confident that I could fit into your team without difficulty.

Closing

Please review my attached resume for a more detailed account of my professional skills and background.

I am available to start working immediately and I look forward to speaking with you at your convenience.

Thank you for your consideration.

Best regards,
Signature of Sender/Sender's Name Printed

E-mail 中译   写信▼   删除   回复▼   寄件者：Ginny

您好：

开头

    为了应聘贵公司于5月15日刊登于人力资源机构的招聘职位，随信附上我的个人简历。

本体

    如您在我的简历中所见，我在零售业务以及服务业有丰富的工作经验，这使我能够与不同类型的顾客打交道，并具有处理投诉的能力。除此之外，我具备完成业务目标的能力。再加上我很灵活，能够适应不同的情况与新环境，我有自信能很快就融入您的团队。

结尾

    有关我更详细的专业技能及背景，请参阅随信附上的个人简历。一旦录用，我可以立即开始上班，很期待能在您方便的时间与您面谈。感谢您的关注。

敬祝安康，
签名档/署名

三、各段落超实用句型

说明：画下划线部分的单词可按照个人情况自行替换。

（一）开头 Introduction

❶I learned from the advertisement you posted on the 104 Manpower Agency on September 24 that you currently have a vacancy for sales manager.

我从贵公司于9月24日刊登在104人力资源机构的广告中得知，贵公司目前有个销售经理的职缺。

❷I saw your advertisement for an office secretary and would like to express my interest.

我看到你们招聘办公室秘书的广告，于是想要表达我对此职务的兴趣。

❸I am writing this letter to express my interest in the position of sales assistant.

谨以此信表达我对销售助理一职的兴趣。

❹This letter is an expression of my keen interest in the vacancy for a maintenance engineer which you advertised on 1234 Manpower Website on November 12.

谨以此信表达我对贵公司于11月12日刊登在1234人力网站上，招聘维修工程师一职的强烈兴趣。

❺With this cover letter, I would like to let you know that I am interested in offering my service as a purchasing assistant in your company.

借此应聘信函，我想让您知道我对贵公司的采购助理一职非常感兴趣。

（二）本体 Body

❶I have three years of working experience in magazine editing.

我在杂志编辑方面有三年的工作经验。

❷My last job was with Jacob's Publishing House as a chief editor in charge of leading a team in the Editorial Department.

我的上一份工作是在雅各出版社担任主编，负责在编辑部门领导一个团队。

❸I had worked as an intern engineer with Huawei for two years, which has given me professional and practical experience in the field of process integration.

我曾在华为担任实习工程师两年，累积了流程集成领域的专业和实践经验。

❹My sufficient experience in material editing would allow me to make an immediate contribution to your organization.

我在教材编辑方面的充足经验将使我能为贵公司立即做出贡献。

❺For the past two years I have worked for a well-known trading company where I gained a reputation for meeting sales targets.

过去两年我任职于一家知名的贸易公司，因完成销售指标而受到好评。

（三）结尾 Closing

❶I can assure that I will contribute to the benefit of your company with my three-year experience in product marketing.

我相信以我在产品营销三年的工作经验，一定能为贵公司贡献收益。

❷I believe that my experience and professional skills would qualify me for this position.

我相信我的经验以及专业技能使我具备胜任此职务的资格。

❸The attached resume can provide you with more details of my professional background.

附件的个人简历能为您提供有关我专业背景的更多细节。

❹Thank you for your time and consideration. I hope to have the opportunity to talk about the position with you in person.

感谢您拨冗考虑。希望能有机会亲自与您讨论此职务。

❺Please see the attached resume for more details of my work experience.

关于我工作经历的更多详情，请参考附件中的简历。

Section Six　Culture Tips

电气元件

电气元件种类繁多，每一类元件都对应着不同的功能，常用电气元件根据功能可分为以下几类。

一、控制元件

1. 接触器。

功能：交流接触器是用来频繁地接通和断开带有负载的主电路或大容量控制电路的电器。

2. 控制用继电器。

功能：用来传递信号或同时控制多个电路，也可以直接用来控制小容量电动机与其他电子执行元件。

3. 主令、按钮、指示灯。

功能：自动控制系统中用于发送控制指令或显示状态的电器。根据不同用途，可分为主令控制器、按钮、转换开关、指示灯、蜂鸣器、带灯按钮等。

二、检测类元件

电流互感器；电流表、电压表、电度表等检测仪表；计时器、计数器等。

三、驱动器及 PLC 系统

变频器、软启动器、多传动变频器直流驱动器、定子调压驱动器、PLC 系统。

四、其他

刀开关、倒顺开关；断路器；熔断器；刀熔开关；过电压保护器（浪涌保护器）；热继电器；其他保护继电器（相序继电器、过压、欠压继电器、过流、欠流继电器、剩余电流继电器等）；变压器；电抗器（进线电抗器、出线电抗器）；滤波器；电阻；接线端子；电线、电缆。

Section Seven Self-evaluation

Rate your learning outcomes in this unit.

Evaluation Grades		Items				
		A	B	C	D	E
Attitude	I can take the initiative to preview before class.					
	I can take an active part in class activities.					
	I can finish tasks carefully and independently.					
Knowledge	I can make good use of words and expressions concerning electrical engineering.					
Ability	I can read and understand the reading material.					
	I can describe an electrician's responsibilities.					
	I can write a simple cover letter.					
Quality	I can understand the significance of being an electrical engineer.					
	I will be a qualified electrician or electrical engineer.					
Is there any improvement over the last unit?		YES		NO		

有志者事竟成。　　　　　　　——《后汉书》
Where there is a will, there is a way.　　—*Book of Later Han*

Unit Two

CAD & CAM

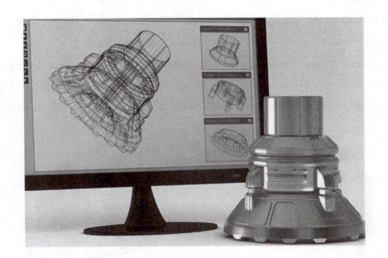

Focus

Section One	Warming Up
Section Two	Dialogue: Introduction to CAD and CAM
Section Three	Passage: Applications of CAD Technology
Section Four	Translation Skills
Section Five	Writing: A Company Introduction Letter
Section Six	Culture Tips
Section Seven	Self-evaluation

Learning Objectives

Upon completion of the unit, students will be able to:

1. Have basic knowledge of CAD and CAM.
2. Fully understand the applications of CAD technology.
3. Keep in mind the advantages of CAD technology.

Section One Warming Up

Group Work

Reorder the following CAD working procedure according to the picture.

A. Work on a draft of the product and draw pictures on a computer.

B. Make a presentation to the client and get feedback from him.

C. Receive a client and get information of his requirements for the new product.

D. Discuss with colleagues for new ideas and revise the draft to meet the client's demand.

E. Refer to some books on designing and collect relevant materials about the other designs of the product.

Unit Two CAD & CAM

Warming Up 听力

Listen to the following material and find out the last step in the CAD working procedure.

Section Two　　Dialogue: Introduction to CAD and CAM

Listen and role-play the following dialogue.

Sally: What are you doing on the computer?

Yang Kang: I'm learning how to use the CAD software. Computer aided design is the cross technology of product design which is applied in engineering field by electronic computer technology.

CAD 和 CAM 简介

Sally: Oh, I've heard about CAD. And what's the difference between CAD and CAM?

Yang Kang: CAD focuses on the design of a product or part, how it looks and how it **functions**. CAM focuses on how to make it. The start of every engineering process begins in the world of CAD.

Engineers make either a 2D or 3D drawing, whether that's a **crankshaft** for an **automobile**, the inner **skeleton** of a kitchen **faucet**, or the hidden electronics in a circuit board. CAD/CAM technology is the result of decades of efforts by **numerous** people in the name of production **automation**. It is the **vision** of **innovators**, inventors, **mathematicians** and **machinists**, who are all working to build the future and drive production with technology.

Sally: In which **professions** is CAD software needed?

Yang Kang: CAD software is used by engineering **disciplines** across all industries. If you're a designer, **drafter**, **architect**, or engineer, you've probably used CAD programs.

Sally: Is there any free CAD software **available** for beginners?
Yang Kang: Certainly.

Words & Expressions

function	[ˈfʌŋkʃn]	vi. 运行；起作用 n. 功能
crankshaft	[ˈkræŋkʃæft]	n. 曲轴
automobile	[ˈɔːtəməbiːl]	n. 汽车 adj. 汽车的
skeleton	[ˈskelɪtn]	n. 骨架；纲要；骨骼 adj. 骨骼的
faucet	[ˈfɔːsɪt]	n. 水龙头
numerous	[ˈnjuːmərəs]	adj. 众多的；许多的
automation	[ˌɔːtəˈmeɪʃn]	n. 自动化
vision	[ˈvɪʒn]	n. 视力；眼光；想象力 v. 幻想；设想
innovator	[ˈɪnəveɪtə]	n. 创新者；改革者
mathematician	[ˌmæθəməˈtɪʃn]	n. 数学家
machinist	[məˈʃiːnɪst]	n. 机械师
profession	[prəˈfeʃn]	n. 行业；职业
discipline	[ˈdɪsəplɪn]	n. 纪律；训练；学科 vt. 训练；惩罚
drafter	[ˈdræftə]	n. 绘图员；起草者
architect	[ˈɑːkɪtekt]	n. 建筑师
available	[əˈveɪləbl]	adj. 可获得的；可购得的；有空的

CAD	（computer aided design）计算机辅助设计
CAM	（computer aided manufacturing）计算机辅助制造
3D drawing	三维制图
circuit board	电路板

 Find Information

Task Ⅰ. Read the dialogue carefully and then answer the following questions.

1. What is Yang Kang doing?

2. Is CAD a cross technology?

3. What's the difference between CAD and CAM?

Unit Two CAD & CAM

4. What can be drawn by engineers on the computer with CAD software?
_____

5. Which professions is CAD software needed in?
_____

Task Ⅱ. Read the dialogue carefully and then decide whether the following statements are true (T) or false (F).

(　　) 1. The central and essential ingredient of CAD/CAM is the electronic computer.

(　　) 2. CAD/CAM technology is the result of several hundred years of efforts by numerous people.

(　　) 3. Many professionals are all working to build the future and drive production with technology.

(　　) 4. If you're an architect, or engineer, you've probably used CAD programs.

(　　) 5. There isn't any free CAD software available for beginners.

Cheer up Your Ears

Listen and write down what you've heard. Then read and recite till you can use them fluently.

1. Computer aided design is a kind of _____ technology.
2. CAD is normally used in _____ design.
3. CAD focuses on the design of a product or part, how it looks and how it _____.
4. The start of every engineering _____ begins in the world of CAD.
5. Engineers make either a 2D or 3D _____.
6. CAD/CAM technology is the result of efforts by _____ people.
7. Innovators, inventors, mathematicians and machinists are all working to drive _____ with technology.
8. CAD technology can be applied in many _____.
9. CAD software is used by engineering _____ across all industries.
10. There is some free CAD software _____ for beginners.

Words Building

DIALOGUE CHEER UP EARS

Task Ⅰ. Choose the best answer from the four choices A, B, C and D.

(　　) 1. We're _____ the process of selling our house.
A. in　　　　　　　B. on　　　　　　　C. for　　　　　　　D. under

(　　) 2. The sofa also functions _____ a bed.
A. in　　　　　　　B. on　　　　　　　C. under　　　　　　D. as

(　　) 3. The ticket is _____ on the day of issue only.
A. available　　　　B. numerous　　　C. avoidable　　　　D. applied

(　　) 4. She _____ well disciplined for her orderliness.
A. should have been　　　　　　　　B. could have been
C. would have been　　　　　　　　D. must have been

(　　) 5. _____ the automobile, Americans soon had a freedom of movement previously unknown.
A. In addition to　　B. On the basis of　　C. Thanks to　　　　D. In case of

Task Ⅱ. Fill in each blank with the proper form of the word given.

1. Employers look favourably on _____ who have work experience. (apply)
2. His _____ career spanned 16 years. (profession)
3. This temple is a textbook example of traditional Chinese _____. (architect)
4. The small individual sources of pollution are far _____ than the large ones. (numerous)
5. If you want to _____, failure is part and parcel with that. (innovator)

Task Ⅲ. Match the following English terms with the equivalent Chinese items.

A. automobile industry　　　　　　1. (　　) 杜比视觉
B. automation production line　　　2. (　　) 电路板
C. Dolby vision　　　　　　　　　3. (　　) 3D 打印机
D. CNC machinist　　　　　　　　4. (　　) 计算机辅助工程
E. senior architect　　　　　　　　5. (　　) 汽车工业
F. 3D drawing　　　　　　　　　　6. (　　) 高级建筑师
G. 3D printer　　　　　　　　　　7. (　　) 三维制图
H. circuit board　　　　　　　　　8. (　　) 护理行业
I. computer aided engineering　　　9. (　　) 数控机械师
J. caring profession　　　　　　　10. (　　) 自动化生产线

 Table Talk

Pair Work

Role-play a conversation about drafters with your partner.

Situation

Imagine you want to be a new drafter in the company. You are interviewed by the interviewer. Now you are talking with him.

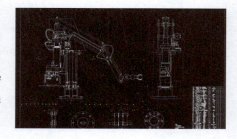

Unit Two CAD & CAM

Section Three Passage: Application of CAD Technology

Pre-reading Questions

1. What do you know about CAD technology?
2. Are you proficient in CAD software?
3. What can you benefit from CAD technology as a drafter?

CAD 技术的应用

  CAD technology is used across many different industries and occupations, and can be used to make **architectural** designs and industrial designs, **generate assembly** and engineering drawings for manufacturing. 3D printing, **verification** and **validation** of design are possible in CAD software.

  Prior to the **advent** of computer aided design, designs needed to be **manually** drawn. Every **object**, line or **curve** needed to be drawn by hand. **Calculations** needed to be done **manually** by an engineer or designer, a very **time-consuming** and **error-prone** process.

  CAD technology changed all of this. CAD has a variety of advantages over manual drawings, which has made it absolutely essential nowadays. Designs can be created and edited in much less time, as well as saved for future use. CAD drawings are not limited to the 2D space of a piece of paper, and can be viewed from many different **angles** to ensure proper fit and design. Calculations are performed by the computer, making it much easier to test the **viability** of designs. Designs can be shared and **collaborated** on in real time, greatly decreasing the **overall** time needed to complete a drawing.

  As CAD programs become more advanced, they will open up more possibilities to designers and engineers and become even more important to an **ever-increasing** range of industries.

Words & Expressions

architectural	[ˌɑːkɪˈtektʃərəl]	adj. 建筑的；建筑学的
generate	[ˈdʒenəreɪt]	vt. 产生；发生；引起
assembly	[əˈsembli]	n. 集会；议会；装配
verification	[ˌverɪfɪˈkeɪʃn]	n. 确认；查证；作证
validation	[ˌvælɪˈdeɪʃn]	n. 确认；批准；验证
advent	[ˈædvent]	n. 出现；到来
manually	[ˈmænjuəli]	adv. 手动地
object	[ˈɔbdʒɪkt]	n. 物体；目标；对象 v. 反对
curve	[kəːv]	n. 曲线
manually	[ˈmænjuəli]	adv. 手动；手动地；人工地

calculation	[ˌkælkjuˈleɪʃn]	n. 计算；估计；深思熟虑
time-consuming	[taɪm kənˈsuːmɪŋ]	adj. 耗费时间的
error-prone	[ˈerəˈprəʊn]	adj. 易错的
angle	[ˈæŋgl]	n. 角度；角；观点
viability	[ˌvaɪəˈbɪləti]	n. 可行性；生存能力
collaborate	[kəˈlæbəreɪt]	vi. 合作
overall	[ˌəʊvəˈɔːl]	adj. 全部的；总体的
ever-increasing	[ˈevəɪnˈkriːsɪŋ]	adj. 不断增长的

prior to...   在……之前

GO Find Information

Task Ⅰ. Read the passage carefully and then answer the following questions.

1. Has CAD technology be widely used in various fields?

2. What is the disadvantage of manual drafting?

3. What is the superiority of CAD technology in terms of designing?

4. What can CAD drawings ensure when they are viewed from many different angles?

5. Will there be more possibilities of CAD technology in the future?

Task Ⅱ. Read the passage carefully and then decide whether the following statements are true (T) or false (F).

(　　) 1. CAD technology can be applied in many different industries and occupations.

(　　) 2. All the architects and industrial designers must be skilled at using CAD technology.

(　　) 3. Drawings can't be finished by hand anymore.

(　　) 4. By sharing and collaborating designs on in real time, drawings can be completed more efficiently.

(　　) 5. The significance of CAD programs is great in all the industries.

Unit Two CAD & CAM

Cheer up Your Ears

Listen and write down what you've heard. Then read and recite till you can use them fluently.

1. CAD technology is used across many different industries and _____.

2. CAD can be used to generate _____ and engineering drawings for manufacturing.

3. 3D _____, verification and validation of design are possible in CAD software.

4. _____ to the advent of computer aided design, designs needed to be manually drawn.

5. Every _____, line or curve needed to be drawn by hand.

6. _____ needed to be done manually by an engineer or designer, a very time-consuming and error-prone process.

7. CAD has a variety of advantages over manual drawings, which has made it absolutely _____ nowadays.

8. Designs can be _____ and edited in much less time, as well as saved for future use.

9. CAD drawings can be _____ from many different angles to ensure proper fit and design.

10. As CAD programs become more advanced, they will become more important to an _____ range of industries.

Words Building

PASSAGE CHEER UP EARS

Task Ⅰ. Translate the following phrases into English or Chinese.

1. CAD 技术的应用_____
2. 建筑设计_____
3. 创建和编辑_____
4. 手工绘图_____
5. CAD 图纸_____
6. prior to... _____
7. share and collaborate _____
8. an ever-increasing range of industries _____
9. a time-consuming process _____
10. real time _____

Task Ⅱ. Choose the best answer from the four choices A, B, C and D.

(   ) 1. Please check in half an hour prior _____ departure.
A. to   B. for   C. in   D. at

(   ) 2. The lawyer _____ a contract for the deal.
A. has drawn   B. draws for   C. drew down   D. drew up

(   ) 3. It is difficult to cross the desert by car, but not _____ impossible.
A. easily   B. greatly
C. absolutely   D. manually

(   ) 4. Once you start eating in a healthier way, weight control will become _____.
A. much easier   B. more easier   C. much more easier   D. more easy

(   ) 5. The population of the country had decreased _____ about 600,000.
A. over   B. by   C. down   D. for

Task Ⅲ. Fill in each blank with the proper form of the word given.

1. The job _____ must have good verbal skills. (apply)

2. _____ rather than saving has become the central feature of contemporary societies. (consume)

3. There were 100 _____ of Hamlet. (perform)

4. The ski school coaches beginners, intermediates, and _____ skiers. (advance)

5. It is essential that he _____ there at once. (send)

Task Ⅳ. Complete the sentences below with the correct form of the words and phrases in the box.

| architectural | assembly | verification | advent |
| object | calculation | angle | viability |

1. Each component is carefully checked before _____.

2. They _____ to putting off the meeting.

3. Finally I'll discuss the _____ of the project.

4. New _____ styles can exist perfectly well alongside an older style.

5. With the _____ of spring, everything comes to life again.

6. The rising sun is especially beautiful to look at from this _____.

7. Please come to the police station with me for identity _____.

8. By his _____, we will be there in an hour.

Task Ⅴ. Translate the following sentences into Chinese.

1. Prior to the advent of computer aided design, designs needed to be manually drawn.

Unit Two　CAD & CAM

2. CAD has a variety of advantages over manual drawings, which has made it absolutely essential nowadays.

_____

3. CAD drawings are not limited to the 2D space of a piece of paper, and can be viewed from many different angles to ensure proper fit and design.

_____

4. Designs can be shared and collaborated on in real time, greatly decreasing the overall time needed to complete a drawing.

_____

5. As CAD programs become more advanced, they will open up more possibilities to designers and engineers and become even more important to an ever-increasing range of industries.

_____

 **Listening**

Task Ⅰ. Listen to 5 short dialogues and choose the best answer.

1. A. The coach broke down on the way.　　B. The coach will arrive soon.
　 C. The tour guide has already arrived.　　D. The tour guide has contacted the driver.

2. A. The sales data.　　B. The meeting agenda.
　 C. The price list.　　D. The travel budget.

3. A. Cancel a reservation.　　B. Change a room for him.
　 C. Give him a morning call.　　D. Book a taxi.

4. A. She will go on a business trip.　　B. She has to finish her report.
　 C. She will visit her parents.　　D. She has to meet her clients.

5. A. To book a table.　　B. To change a flight.
　 C. To make a complaint.　　D. To place an order.

A 级听力单选

A 级听力填空

Task Ⅱ. Listen to the passage and fill in the blanks with the missing words or phrases.

Good morning, everybody. Today, I'd like to introduce you to our tour for tea lovers. As you know, tea is an 1._____ of Chinese tradition. You may have no idea about how the tea grows and how it is made. Our tour will enable you to 2._____ the tea culture in China. Hangzhou is the 3._____ of Longjing tea, which is one of the most famous green teas in China. During this tour, you will have the chance to go to a tea farm, 4._____ tea leaves, visit a tea farmer's house, and learn the art of tea-making and 5._____ of Longjing tea. I hope you're pleased to travel with us to learn more about Chinese tea culture.

31

 **Extensive Reading**

Directions: *After reading the passage, you are required to choose the best answer from the four choices for each statement.*

What's Next for CAD/CAM Technology?

The following trends may show us where the next great leap in CAD/CAM technology will emerge.

Artificial Intelligence: Incorporating AI into design software allows the automation of design tasks, enhances quality control by anticipating design errors and (with machine learning) paves the way for the creation of unique designs without human input.

Cloud Collaboration: Cloud technology allows CAD/CAM to move beyond a single computer at a workplace to universal access through a SaaS (software-as-a-service) model. This will mean several people can work on the same project at once while sharing across departments and geographies has become much easier.

Virtual Reality: VR helmets and VR glasses can be used to take advantage of the life-like visualization offered by sophisticated CAD software. For instance, an architect can now offer a "walkthrough" of a building that exists only as a digital model.

Customization: Software providers are moving away from a one-size-fits-all solution to provide the option of configuring CAD/CAM to suit your work environment, and choose only the tools that will be required for a particular job. This may be a way to offer affordability by cutting out dozens of features that the average user may never need.

1. The next great leap in CAD/CAM technology will emerge in the following trends. _____.

A. AI and VR   B. Customization
C. Cloud collaboration   D. A, B and C

2. The advantages of incorporating AI into design software include the following except _____.

A. designing multifunctional machines for people
B. allowing the automation of design tasks
C. enhancing quality control by anticipating design errors
D. paving the way for the creation of unique designs without human input

3. Cloud technology allows CAD/CAM to move beyond a single computer at a workplace to universal access through a SaaS model. This will mean _____.

A. people can work on the same project at once while sharing across departments and areas has become much easier

  B. several people work on the same project at once while sharing in the same department and geography has become much easier

  C. several people are necessary on the same project at once while sharing in the same department and geography has become much easier

  D. people can work on the different projects at once while sharing across departments and areas has become much easier

  4. How can an architect offer a "walkthrough" of a building that exists only as a digital model? _____ .

  A. By CAD and CAM software   B. By sophisticated CAD software

  C. By CAD or CAM software   D. By sophisticated CAM software

  5. Providing the option of configuring CAD/CAM to suit your work environment, and choosing only the tools that will be required for a particular job is called _____ .

  A. artificial intelligence   B. cloud collaboration

  C. virtual reality   D. customization

Section Four　Translation Skills

科技英语的翻译方法与技巧——增译法

  在科技英语汉译时，词语的增译主要体现在两个方面：第一，根据语义上的需要进行增补，使译文意义更加自然和通畅；第二，根据语法上的需要进行词语增补，使译文更符合汉语的表达习惯。

一、增词翻译

（一）增译名词

1. 抽象名词后增译名词。

  翻译时为了使译文更通顺和清晰，可以根据上下文逻辑关系适当增补一些范畴词，常见的有原理、作用、现象、效应、问题、状况、办法、条件、范围、结果、对象、过程、形式、能力、术语、品质、地位、因素、关系、观点等。如：

  Computer implementation is mandatory but cannot be done entirely by persons who are unfamiliar with the method.

  【译文】计算机执行过程是按照人的指令进行的，但若完全交给不懂方法的人来操作，计算机则无法执行任务。

2. 形容词前增译名词。

  科技英语中有许多描述事物体积、价格、质量、长度、面积、性能、功能等的形容词，若直译的话，会造成语句不通顺，甚至导致误解和曲解等情况的发生。因此，在英译汉的过程中，需要在这样的形容词前面或后面增译相应的名词。如：

The technologies and applications must be easy and relatively cheap.
【译文】这些技术和应用必须操作简单、价格相对实惠。

3. 介词词组等前后增译名词。

科技英语中引用他人的发言、意见、成果或某品牌的产品时，常以人名、地名、成果和品牌名的形式出现，翻译时要在后面增补具体名词。如：

From the surveys, there seem to be two main ways in which to convince executives that connectedness projects are important, hard financial numbers and customer service.
【译文】从这些调查结果看，能够说服行政主管连通项目重要性的途径主要有两条：确凿的财政数据和客户服务。

二、增译动词

用汉语动词补充英语动名词、名词或介词的意义，以使译文准确和通顺。可以增加的动词有进行、实现、获得、引起、出现、发生、限于、遭受、调整、使、制定……。如：

The data can be entered directly into analog-digital converters and computers for processing.
【译文】可以把数据直接输入模拟－数字转换器和计算机中进行处理。

三、增译形容词

英语描述物体的性能、质量、体积等特点的时候比较精简，通常需要通过上下文推断才能了解其内在含义。为了让译文含义更加清楚、直观，翻译时需要在名词前或后增加形容词。如：

Software developers know they want to produce a quality product.
【译文】软件开发人员都知道他们想要研制出一个高质量的产品。

四、增译各类数词和量词

汉语中有三百多个量词，如颗、台、套、张、页、方、间、批、组、条、件等。相比之下，英语中量词要少许多，更多的是数字直接与英文名词连用。所以在英译汉时，为了使描述的事物显得更加丰富多彩，常需要增加量词。如：

a website 一个网址　　　　　　　　a car 一辆车
a video cassette 一盘录影带　　　　a cable 一根网线
a circuit board 一块电路板　　　　two pills 两片药
a machine 一台机器　　　　　　　three batteries 三节电池

英语有连续列举的情况，那么在汉译时要根据上下文增加"两个、三点、四类、五方面、等几个方面、几种情况"等概念性的词语，这样可以对原文进行恰当地概括，读起来容易有一个整体的感觉，也使得译文内容更醒目。如：

A hybrid four-wheel-drive transmission, a convertible top, and a removable windshield made it the most versatile smart car.

【译文】油电混合四轮驱动传输、敞篷车顶和可移动挡风玻璃这三大特点使它成为最通用的智能车。

五、语法增补翻译

（一）增译曲折变化词

1. 表述复数概念的词。

翻译有关人或物的英语名词时，可以在复数概念的名词前增加形容词（如：许多、很多）、量词（如：百、千、万）、物量词（如：几个、若干、一批、各种）、代词（如：每、各、们、双方）、重叠词（如：人人、重重、一列列、种种、层层）和其他汉语中带有复数语义特征的词，如"群、众、多个、数个"等。

The virus may survive weeks and months.

【译文】该病毒可存活数周乃至数月。

2. 表示时态的词。

英语的时态变化可以直接体现在动词的形态变化上。汉语的时态变化则无法通过谓语动词的词形变化来体现，而是要借助增加一些副词，如"正、正在、过、了、着、曾经、已经、一直、将要、会、能"等。因此，科技英语译汉时，需要视具体情况增译这些副词。

Electronic data interchange (EDI) has been extensively and successfully implemented, and is growing in popularity.

【译文】电子数据交换已经得到成功、广泛的应用，并且正在逐渐走向普及。

（二）增加关联词

科技英语汉译时，需要增加合适的关联词，使语句之间的关系更加合乎逻辑，语义更清晰。

1. 如果分词短语或独立分词结构含有时间、原因、条件、让步等状语意义，翻译时可增加"当……时，……之后，因为……，如果……，假若……，虽然……但……"等词。

Without the cooling and lubrication systems, an engine would be badly damaged by heat and friction within minutes.

【译文】如果没有冷却和润滑系统，发动机将在几分钟内因为发热和摩擦而严重受损。

2. 当动词不定式和不定式短语表示目的或结果状语时，通常可以增加"为了、要、以便、就、从而、结果"等词。如：

An engine must have an adequate supply of engine oil at all times to operate without damage to internal parts.

【译文】为了保证发动机运行且内部零件不被损坏,必须始终给发动机提供充足的机油。

3. 英语倒装语序表示的虚拟条件状语从句,译成汉语时,往往可增加"如果……便……,假如……就……,万一……就……",以及"只要……的话,就……"等连词。如:

Had computers not been given this flexibility, it is probable that they would not have met with such widespread use.

【译文】如果计算机不是如此方便灵活,它们就不可能被如此广泛地应用。

(三) 被动句的增译

科技英语被动语态使用频繁,汉译时常常要转换成主动语态。当英语被动语态句中由表示"知道、了解、看见、认为、发现、考虑、使用"等意义的动词充当谓语时,转换成主动语态句后,需要增补原文中没有出现的行为主体,如"人们、我们、大家、有人"等词。如:

Bioactive antioxidant substances were found in fruits and vegetables.

【译文】人们发现水果和蔬菜中存在生物活性抗氧化物质。

(四) 增加祈使句中的礼貌词

在英语产品说明书、宣传册和广告用语中,常使用 be sure to、be careful、make sure、remember、note that、do 等类似的用语,表示命令、指示、警告、请求或叮嘱等含义。在汉译时,常常根据对象、场合增译诸如"请、要、切、应、须、得、千万、一定、务必、注意、必须、应当"等具有命令、提示、指示、警告等语气的词。如:

Make sure you have the correct tools, parts, and other materials before you begin to work.

【译文】在开始工作前,应确保配备合适的工具、零部件和其他材料。

Section Five Writing: A Company Introduction Letter

向厂商介绍公司

一、写作要点 Key Points

Step 1: 说明来信的目的是介绍公司。
Step 2: 简短陈述公司的主要运营项目。
Step 3: 主动提出期盼未来能够合作。

二、实际 E-mail 范例

| 写信▼ | 删除 | 回复▼ | 寄件者： | Ginny |

Dear Mr. /Ms. /Dr. /Professor Surname,

Introduction
I would like to take this opportunity to introduce our company to you. JSB Trading Ltd. has been engaged in import and export business for the last 15 years.

Body
We specialize in the import and export of household appliances among Europe, America and Asia, and we have gained a fine reputation for our excellent services.

Closing
We would like to offer our services to your department store. Our marketing manager will contact you shortly and provide you with a detailed description of our services.
We are looking forward to working with you.

Best regards,
Signature of Sender/Sender's Name Printed

E-mail 中译 | 写信▼ | 删除 | 回复▼ | 寄件者： | Ginny

亲爱的……先生/女士/博士/教授，您好：

开头
　　我希望能借用这个机会向您介绍本公司。JSB 贸易有限公司在过去的 15 年来一直从事进出口业务。

本体
　　我们专门从事欧洲、美洲及亚洲之间的家电进出口业务，并且因为服务卓越而在业界享有极佳的声誉。

结尾
　　我们很希望能为贵百货公司提供服务。我们的营销经理将会很快与您联系，为您提供有关本公司服务项目的详细说明。
　　我们很期待与您合作。

敬祝安康，
签名档/署名

三、各段落超实用句型

说明：画下划线部分的单词可按照个人情况自行替换。

（一）开头 Introduction

❶Please allow me to take this opportunity to introduce our company.
请容许我借此机会向您介绍本公司。

❷I am writing this letter to introduce our company and our services to you.
谨以此信向您介绍本公司以及本公司的服务项目。

❸I am writing to you on behalf of JSB Trading Ltd.
本人谨代表JSB 贸易股份有限公司写这封信给您。

❹Our company has been involved in the event organizing business for the past six years.
过去六年来，本公司一直从事活动策划业务。

❺JSB has been engaged in the export of household appliances since it was established in 2003.
JSB 公司自2003 年成立以来，一直从事家用电器外销的工作。

（二）本体 Body

❶I would like to pay you a visit in order to present the services we provide in detail.
我希望能拜访您，向您详细介绍我们所提供的服务项目。

❷I would like to meet with you in person and offer more detailed information about our products.
我希望能私下与您碰面，以提供有关本公司产品更详细的信息。

❸Our product catalogue is attached for your consideration.
谨随信附上本公司产品目录，供您参考。

❹Our sales specialist, Mr. Wu, will get in touch with you by the end of this week.
我们的销售专员吴先生会在本周内与您联系。

❺We provide high quality service for reasonable price.
我们以合理的价格，提供高质量的服务。

（三）结尾 Closing

❶Please feel free to contact me if you have any questions regarding our services.
若您对本公司的服务项目有任何问题，欢迎与我联络。

❷Please let me know if you need any further information regarding our products.
若您需要有关本公司产品的进一步信息，请与我联系。

❸Product samples will be sent to you on demand.
索取后产品样本将会寄送给您。

❹We would like to provide the services of our company to your restaurant.
我们希望能为贵餐厅提供本公司的服务。

❺We look forward to working with you in the near future.
我们期待能在不久的将来与您合作。

Section Six　Culture Tips

中国 CAD 软件的发展历程

　　CAD 技术使产品的设计制造和组织生产的传统模式产生了深刻的变革，成为产品更新换代的关键，被人们称为产业革命的发动机。我国自 20 世纪 60 年代开始研发 CAD 软件，经过"六五""七五""八五"的成果积累，CAD 软件产业在建筑、石化、轻工、造船、机械、航空、航天、电子等行业，以及二维参数化绘图、有限元分析、曲面造型和数控加工编程等方面都已经初具规模，成绩显著。

　　国产 CAD 发展较晚，较之国际落后较大，实际发展年限基本自 20 世纪 90 年代开始。此时国际市场 CAD 已开始在设计领域全面普及。早期受限于国内整体计算机水平，国产 CAD 发展缓慢，21 世纪国际 CAD 龙头进入国内市场，迅速占据主要市场份额，3D 份额占比持续提升。随着国内技术发展，中望软件等国产企业持续突破，2D 已占据国内 20% 左右的市场份额，预计未来在 2D 和 3D 领域市场份额将持续提升。

　　2021 年，我国制造业 CAD CR3 国际企业合计占比 55.7%，其中达索占比最高，为 25.3%。国产中望软件、苏州浩辰和华天软件等占比整体表现为持续提升，整体国产率接近 20% 左右，其中中望软件发展最快市场份额占比达 11.4%。从 3D CAD 软件来看，在技术要求更高的背景下，整体市场集中度更高，占比最高的达索市场份额为 31.7%，国产化率整体更低，中望软件虽然国内份额排名仍在第四位，但市场份额占比下降至 3.9%。随着国产 CAD 的持续发展，3D CAD 市场仍有较大的渗透空间。

Section Seven Self-evaluation

Rate your learning outcomes in this unit.

Evaluation Grades		Items				
		A	B	C	D	E
Attitude	I can take the initiative to preview before class.					
	I can take an active part in class activities.					
	I can finish tasks carefully and independently.					
Knowledge	I can make good use of words and expressions concerning CAD.					
Ability	I can read and understand the reading material.					
	I can describe the advantages of CAD technology.					
	I can write a company introduction letter.					
Quality	I can understand the importance of CAD technology.					
	I can talk about CAD software with others.					
Is there any improvement over the last unit?		YES			NO	

工欲善其事，必先利其器。 ——《论语》
A handy tool makes a handy man. —*The Analects of confucious*

Unit Three

Automation

Focus

Section One	Warming Up
Section Two	Dialogue: Visiting an Automated Factory
Section Three	Passage: Advantages and Disadvantages of Automation
Section Four	Translation Skills
Section Five	Writing: An Event Invitation Letter
Section Six	Culture Tips
Section Seven	Self-evaluation

Learning Objectives

Upon completion of the unit, students will be able to:

1. Have basic knowledge of automation, including the advantages and disadvantages.
2. Have an understanding of the working process in an automated factory.
3. Know some important information in an automated factory.

Section One   Warming Up

Warming Up 听力

Group Work

Match the areas that automation is widely used.

A. Medical treatment.   B. Industry.   C. Military.
D. Family.   E. Transport.   F. Agriculture.

Listen to the dialogue and find out what Yang Kang is doing now.

Unit Three   Automation

Section Two — Dialogue: Visiting an Automated Factory

Listen and role-play the following dialogue.

参观自动化工厂

Sally: Good morning, Mr. Yang.

Yang Kang: Good morning, Ms. Wood.

Sally: Welcome to our factory.

Yang Kang: Thank you. I've been looking forward to visiting your factory.

Sally: I'll show you around and explain the **operation** as we go along.

Yang Kang: That'll be most helpful. Is the factory **fully-automated**?

Sally: Not completely. Our **production process** is **partially-automated**. We use robots on the **production line** for **routine assembly** jobs but some of the work is still done **manually**.

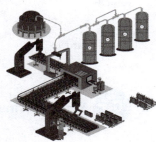

Yang Kang: What about **supply** of parts to the production line?

Sally: Well, the parts are **automatically** selected from the store room using a **bar-code** system. And there is an **automatic feeder** which takes them to the **conveyor belt** at the start of the production line.

Yang Kang: What about the smaller **components**?

Sally: They're **transported** to the **workstations** on automated **vehicles**—robot trucks—which run on guide rails around the factory.

Yang Kang: That's wonderful.

Sally: Yes. Shall we take a break now?

Yang Kang: Okay.

Words & Expressions

operation	[ˌɒpəˈreɪʃn]	n. 操作，运转
fully-automated	[ˈfʊliːˈɔːtəmeɪtɪd]	adj. 全自动化的
partially-automated	[ˈpɑːʃəliˈɔːtəmeɪtɪd]	adj. 部分自动化的
production	[prəˈdʌkʃn]	n. 生产，制作；产量
process	[ˈprəʊses]	n. 过程，进程
routine	[ruːˈtiːn]	n. 例行公事，惯例

assembly	[əˈsembli]	n. 装配，组装
manually	[ˈmænjuəli]	adv. by hand 手动的
supply	[səˈplaɪ]	n. 供给，补给
automatically	[ˌɔːtəˈmætɪkli]	adv. 自动地
bar-code	[ˈbɑːr kˈəʊd]	n. 条形码
automatic	[ˌɔːtəˈmætɪk]	adj. 自动的
feeder	[ˈfiːdə(r)]	n. 送料机
conveyor	[kənˈveɪə]	n. 输送机
belt	[belt]	n. 传送带
component	[kəmˈpəʊnənt]	n. 部件，元件
transport	[ˈtrænspɔːt]	vt. 运送
workstation	[ˈwɜːksteɪʃn]	n. 工作站
vehicle	[ˈviːɪkl]	n. 交通工具，车辆

look forward to　　　　　　　　　　　　　　展望，期待
show...around　　　　　　　　　　　　　　带领……参观（某地）

 **Find Information**

Task Ⅰ. Read the dialogue carefully and then answer the following questions.

1. Who will show Yang Kang around the factory?

2. Is the factory fully-automated?

3. What are robots used for?

4. What about supply of parts to the production line?

5. What about the smaller components?

Task Ⅱ. Read the dialogue carefully and then decide whether the following statements are true (T) or false (F).

(　　) 1. Yang Kang is visiting a factory accompanied by Sally.

(　　) 2. The factory is fully-automated.

(　　) 3. Some of the work is still done manually in the partially-automated factory.

(　　) 4. Automation is inseparable from robotics.

(　　) 5. Sally's explanation is very helpful to Yang Kang.

Unit Three Automation

Cheer up Your Ears

Listen and write down what you've heard. Then read and recite till you can use them fluently.

1. I've been looking forward to _____ your factory.
2. I'll show you around and explain the _____ as we go along.
3. Is the factory _____-automated?
4. Our production process is _____-automated.
5. We use _____ on the production line for routine assembly jobs but some of the work is still done manually.
6. What about _____ of parts to the production line?
7. The parts are automatically selected from the store room using a _____ system.
8. There is an automatic _____ which takes them to the conveyor belt at the start of the production line.
9. What about the smaller _____?
10. They're transported to the _____ on automated vehicles—robot trucks, which run on guide rails around the factory.

Words Building

DIALOGUE CHEER UP EARS

Task Ⅰ. Choose the best answer from the four choices A, B, C and D.

(   ) 1. We look forward to _____ you as a member.
A. have   B. having   C. be having   D. has

(   ) 2. Yesterday we had the pleasure of visiting an automated factory, _____ products are well sold at home and abroad.
A. which   B. that   C. whose   D. where

(   ) 3. Last week two engineers _____ to help solve the technical problem of the project.
A. have sent   B. were sent   C. sent   D. had sent

(   ) 4. It has been quite some time since we _____ the factory.
A. visit   B. have visited   C. had visited   D. visited

(   ) 5. Something must _____ to improve the work efficiency of employees.
A. be done   B. do   C. have done   D. to be done

Task Ⅱ. Fill in each blank with the proper form of the word given.

1. The expert made a very _____ suggestion for the project. (help)
2. In the last ten years, _____ has reduced the work force here by half. (automate)

3. This manual must be provided for the machine _____. (operate)

4. The _____ of the machine is very simple. (operate)

5. The most important part of this _____ line is, of course, the programmer. (produce)

Task Ⅲ. Match the following English terms with the equivalent Chinese items.

A. automated factory   1. (　　) 自动柜员机
B. automated system   2. (　　) 计算机自动设计
C. automated welding   3. (　　) 自动化测试
D. automated teller machine   4. (　　) 全自动化的
E. automated testing   5. (　　) 自动化工厂
F. computer-automated design   6. (　　) 自动化施工
G. automated construction   7. (　　) 半自动化的
H. automatic feeder   8. (　　) 自动化系统
I. fully-automated   9. (　　) 自动送料机
J. partially-automated   10. (　　) 自动化焊接

 **Table Talk**

Pair Work

Role-play a conversation about automation with your partner.

Situation

Discuss the impact of automaton on your own life and lists its main advantages and disadvantages. Now you are talking with your partner.

Section Three Passage: Advantages and Disadvantages of Automation

Pre-reading Questions

1. Can you say something about automated factories?
2. Do you know the advantages and disadvantages of automated factories?

自动控制系统

A further development of **mechanization** is **represented** by automation, which means the use of control system and information technologies to reduce the need for both **physical** and **mental** work to produce goods.

Unit Three Automation

Automation has had a great **impact** on industries over the last century, changing the world **economy** from industrial jobs to service jobs.

The following sums up the main advantages and disadvantages of automation.

Advantages:

Speeding up the development **process** of society;

Replacing human operators in tasks that include hard physical or **monotonous** work;

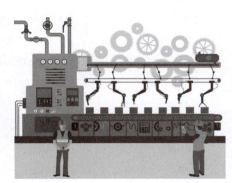

Saving time and money as human operators can be employed in higher-level work;

Replacing human operators in tasks done in dangerous environments (fire, space, **volcanoes**, **nuclear facilities**, underwater);

Higher **reliability** and **precision** in performing tasks;

Economy **improvement** and higher **productivity**.

Disadvantages:

Disastrous effects on the environment (pollution, traffic, energy **consumption**);

Sharp increase in unemployment rate due to machines replacing human being;

Technical **limitations** as current technology is unable to automate all the desired tasks;

Safety **threats** as an automated system may have a limited level of intelligence and can make errors;

High **initial costs** as the automation of a new product requires a large initial **investment**.

Automation has both advantages and disadvantages, but as a whole, the advantages of the rapid development of automation still far **outweigh** the disadvantages.

Words & Expressions

advantage	[əd'vɑːntɪdʒ]	n. 有利条件，优势，优点
disadvantage	[ˌdɪsəd'vɑːntɪdʒ]	n. 不利条件，不利
mechanization	[ˌmekənaɪ'zeɪʃn]	n. 机械化
represent	[ˌreprɪ'zent]	vt. 代表；象征
physical	['fɪzɪkl]	adj. 身体的，肉体的
mental	['mentl]	adj. 精神的，头脑的，心理的
impact	['ɪmpækt]	n. 影响；撞击；冲击力；强大作用
economy	[ɪ'kɒnəmi]	n. 经济体制，经济状况
process	['prəʊses]	n. 过程，进程
monotonous	[mə'nɒtənəs]	adj. 单调乏味的
volcano	[vɒl'keɪnəʊ]	n. 火山

nuclear	[ˈnjuːkliə(r)]	adj. 核的，原子核的
facility	[fəˈsɪləti]	n. 设备，设施
reliability	[rɪˌlaɪəˈbɪləti]	n. 可靠性
precision	[prɪˈsɪʒn]	n. 精确，准确
improvement	[ɪmˈpruːvmənt]	n. 改进，改善
productivity	[ˌprɒdʌkˈtɪvəti]	n. 生产率，生产力
disastrous	[dɪˈzɑːstrəs]	adj. 灾难性的，造成灾害的
consumption	[kənˈsʌmpʃn]	n. 消费，消耗；消费〔耗〕量
limitation	[ˌlɪmɪˈteɪʃn]	n. 限制，限定
threat	[θret]	n. 威胁，恐吓
intelligence	[ɪnˈtelɪdʒəns]	n. 智力，智慧；理解力
initial	[ɪˈnɪʃl]	adj. 最初的，开头的
cost	[kɒst]	n. 成本，费用
investment	[ɪnˈvestmənt]	n. 投资
outweigh	[ˌaʊtˈweɪ]	vt. 比……重要；比……有价值

have an impact on...	对于……有影响，有冲击
sum up	总结，概述
speed up	加速；使加速
due to	由于

GO Find Information

Task Ⅰ. Read the passage carefully and then answer the following questions.

1. Does automation just reduce the need for physical work to produce goods?

2. Why has automation had a huge impact on industries over the last century?

3. Automation can replace human operators in tasks done in dangerous environments. Can you give us some examples?

4. Is the current automation technology able to automate all the desired tasks?

5. What should you do if any accident occurs?

Unit Three Automation

Task Ⅱ. Read the passage carefully and then decide whether the following statements are true（T）or false（F）.

(　　) 1. There aren't any disadvantages to automation.
(　　) 2. Robots can replace human operators in any tasks.
(　　) 3. Automation makes more people lose their jobs due to machines replacing human beings.
(　　) 4. Automated factory doesn't pollute the environment.
(　　) 5. Automation has had a great impact on our life in the past years.

Cheer up Your Ears

Listen and write down what you've heard. Then read and recite till you can use them fluently.

1. A further development of mechanization is represented by _____.
2. Automation means the use of _____ system and information technologies to _____ the need for both physical and mental work to produce goods.
3. Automation has had a great impact on _____ over the last century, changing the world economy from industrial jobs to service jobs.
4. Speeding up the development process of _____.
5. Replacing human operators in tasks that include hard _____ or monotonous work.
6. Saving _____ and money as human operators can be employed in higher-level work.
7. Replacing human operators in tasks done in _____ environments.
8. Disastrous effects on the _____, including pollution, traffic, energy consumption.
9. Sharp increase in _____ rate due to machines replacing human being.
10. As a whole, the _____ of the rapid development of automation still far outweigh the _____.

Words Building

Task Ⅰ. Translate the following phrases into English or Chinese.

1. 自动化工厂　　_____
2. 控制系统　　　_____
3. 信息技术　　　_____
4. 体力劳动　　　_____
5. 生产线　　　　_____
6. unemployed rate　_____
7. mental work　　_____
8. human operator　_____

PASSAGE CHEER
UP EARS

9. energy consumption　　_____

10. safety threat　　_____

Task Ⅱ. Choose the best answer from the four choices A, B, C and D.

(　　) 1. Great changes _____ in the factory since 10 years ago.

A. take place　　　　　　　　B. are taken place

C. have taken place　　　　　D. have been taken place

(　　) 2. Robots continue to have an impact _____ blue collar jobs.

A. to　　　　　B. on　　　　　C. with　　　　　D. for

(　　) 3. Robots can do repetitive tasks instead of humans, _____ means some people will lose their jobs.

A. where　　　B. who　　　　C. which　　　　D. that

(　　) 4. Her work is showing some signs of _____.

A. improvement　　B. investment　　C. agreement　　D. advertisement

(　　) 5. The breakdown was _____ a mechanical failure.

A. belong to　　B. stick to　　C. lead to　　D. due to

Task Ⅲ. Fill in each blank with the proper form of the word given.

1. This _____ was dismissed for laziness. (employ)

2. The management is doing its best _____ the situation. (improve)

3. With the _____ of modern technology, robots are going into many fields of human beings. (develop)

4. Most people don't realize that they are breathing _____ air. (pollute)

5. The main purpose of _____ is to create wealth. (industrial)

Task Ⅳ. Complete the sentences below with the correct form of the words and phrases in the box.

| show... around | speed up | production line | automated system |
| intelligence | automation | pollute | physical |

1. We've changed from traditional methods of production an _____.

2. Working with tools also helped to develop human _____.

3. When it comes to _____, the chemical industry is a major offender.

4. The factory should take steps to _____ solving the problem.

5. Labor refers to the use of mental or _____ work to produce goods.

6. _____ can free workers from hard, boring work.

7. He became an adjuster on the _____.

8. When you settle down, I'll _____ you _____ and introduce you to the department managers.

Task Ⅴ. Translate the following sentences into Chinese.

1. Automation has had a great impact on industries over the last century, changing the world

economy from industrial jobs to service jobs.

2. Replacing human operators in tasks that include hard physical or monotonous work.

3. Saving time and money as human operators can be employed in higher-level work.

4. Replacing human operators in tasks done in dangerous environments (fire, space, volcanoes, nuclear facilities, underwater).

5. Automation has both advantages and disadvantages, but as a whole, the advantages of the rapid development of automation still far outweigh the disadvantages.

 Listening

A级听力单选

A级听力填空

Task Ⅰ. Listen to 5 short dialogues and choose the best answer.

1. A. Prepare the documents.
   B. Book a room for the meeting.
   C. Ask all the managers to attend the meeting.
   D. Make several copies of the meeting agenda.

2. A. He is going to a party.　　　　　B. He is making a plan.
   C. He likes to go to the cinema.　　D. He can't attend the lecture.

3. A. The job is interesting.　　　　　B. The environment is friendly.
   C. The salary is attractive.　　　　D. The colleagues are nice.

4. A. The labour cost has risen.　　　B. The income tax has gone up.
   C. The management cost is on the rise.　D. The raw material is in short supply.

5. A. Lowering the product price.　　B. Conducting a market survey.
   C. Making a promotion plan.　　　D. Improving the product design.

Task Ⅱ. Listen to the passage and fill in the blanks with the missing words or phrases.

We are now about to close the marketing conference. We owe thanks to every member of staff who made the conference a 1. _____. We would like to thank all of the speakers who have made 2. _____ and impressive speeches. Our thanks also go to every one of you for your contributions and your appreciation. The energy and the enthusiasm surrounding this conference have been 3. _____. And now we'll go back to our business and put those 4. _____ into action. It's time to get back to start working on the next phase. All customers are out there waiting for us to 5. _____.

Extensive Reading

Directions: After reading the passage, you are required to chose the correct answer.

Sensors

Almost every industrial automated process requires the use of sensors and transducers, which are very advanced devices capable of measuring and sensing the environment and translating physical information (e. g. variations of light, pressure, temperature and position) into electrical signals. The sensor picks up the information to be measured and the transducer converts it into electrical signals that can be directly processed by the control unit of a system.

Because of the industrial and scientific importance of measuring, sensors are widely used in a variety of fields, such as medicine, engineering, robotics, biology and manufacturing. Traditional machines have difficulty measuring small differences in product size, so sensors can be particularly useful as they can discriminate down to 0. 000 13 mm. They can also detect temperature, humidity and pressure, acquire data and alter the manufacturing process. Sensors are also vital components of advanced machines, such as robots.

There are two types of sensors, analogue and digital. Analogue sensors operate with data represented by measured voltages or quantities, while digital ones have numeric digital outputs which can be directly transmitted to computers.

The sensors usually employed in manufacturing are classified as mechanical, electrical, magnetic and thermal, but they can also be acoustic, chemical, optical and radiation sensors. Moreover, according to their method of sensing, they can be tactile or visual. Tactile sensors are sensitive to touch, force or pressure and they are used to measure and register the interaction between a contact surface and the environment. These sensors are used in innumerable everyday objects, such as lift buttons and lamps which turn on and off by touching the base. Visual sensors, instead, sense the presence, shape and movement of an object optically. They are becoming more and more important in surveillance systems, environment and disaster monitoring and military applications.

1. Sensors pick up _____ to be measured.

   A. electrical signals   B. physical information   C. the control unit

2. Physical data is translated into electrical signals by _____ .

   A. the transducer   B. the sensor   C. a computer

3. Sensors _____ used to alter the manufacturing process.

   A. can't be   B. are never   C. can be

4. _____ sensors can transmit data directly to computers.

   A. Chemical   B. Digital   C. Analogue

5. Tactile sensors are commonly used in _____ .

   A. everyday objects   B. military applications   C. sophisticated machinery

Section Four Translation Skills

科技英语翻译技巧——省译法

省译绝不等同于省义。省译是省略译文中不需要的词语，省译后并不省略原文的信息，不改变原文的意思。省译大致可以分为两大类型，即语法上的省译和重复内容的省译。

一、语法上的省译

（一）虚词省译

1. 省译冠词。

定冠词 the 的省译主要涉及泛指类别、表示世界上独一无二的事物、位于定语之前、位于形容词比较级和最高级前、位于序数词前和某些固定词组等。如：

High-quality, publicly funded research is the wellspring of medical breakthroughs.

【译文】许多医学领域的突破源于高质量和公共基金资助的研究。（泛指类别）

This is why the Smart Car took Europe by storm.

【译文】这就是斯玛特汽车能迅速进入欧洲市场的原因。（表示世界上独一无二的事物）

A chlorine atom is regenerated in the second reaction.

【译文】第二次化学反应过程又产生了氯原子。（位于序数词前）

Those new areas add to existing branches of chemistry and at the same time create new areas of study.

【译文】那些新的领域扩充了现有的化学分支，同时也形成了新的研究领域。（固定搭配）

当不定冠词 a 和 an 在泛指某一类事物中的任何一个或用于某些固定搭配中时也可以省译。如：

The tyre manufacturers only sell a minority of their output to OEMs.

【译文】一些轮胎制造商只是将少量产品卖给原始设备制造商。（固定搭配）

但是当不定冠词与单数可数名词连用时，并明显地表示数量"一、每一"或"同一"等意思时，往往不宜省译。如：

Every function in JavaScript is represented as an object—more specifically, as an instance of function.

【译文】每一个 JavaScript 函数都被表示为一个对象。更确切地说，它是一个函数实例。

2. 省译介词。

英译汉时，介词一般可以省略不译。如：

The first transistorized computer was demonstrated at the University of Manchester in 1953.

【译文】1953 年，曼彻斯特大学展示了第一台晶体管计算机。（状语中的介词省译）

Chances are that you're already familiar with the mouse and know how the thing works and what to do with the buttons.

【译文】巧的是你已经熟悉鼠标，了解它的工作原理和那些按键的功能。（跟在形容词后面的介词省译）

介词也广泛存在于科技英语专有名词、术语和行话中，如书名、地名、机构、发明等。翻译时，其中的介词往往可以省略不译。如：

Periodic table of elements 元素周期表

Schedule time of departure 预计离港时间

The International Organization for Standardization 国际标准化组织

Cash on delivery 交货付款

3. 省译连词。

英语重形合，句子内部的逻辑关系主要通过连词来建立；汉语重意合，注重意义的连贯。因此，科技英语翻译时，常常可以省译连词。如：

Since the process is a mechanical one and does not require heat, it can be very precisely controlled.

【译文】整个过程是一个机械加工过程，且无须加热，因此可以精确操控。（since 引导的原因状语从句，表示因果关系）

（二）实词省译

1. 省译代词。

科技英语翻译中，代词省译可以使译文更加准确、简洁。如：

When you step on the brake pedal, forces are applied to the hydraulic caliper.

【译文】踩刹车时，力会传到液压制动器上。（第二人称 you 省译）

2. 省译动词。

英语的谓语动词种类繁多，有一类是含义被虚化的动词。针对这类动词词组，汉译时，可以省译虚化动词，直接翻译其所搭配的其他词。如：

Small engines give poor acceleration so to achieve acceptable performance hybrid technology is required.

【译文】小型发动机加速较慢，所以需要使用混合动力来达到理想性能。（虚化动词 give 省译）

3. 省译引导词。

there 和 it 的省译。英语存在 there be（seem, appear, exist, happen, stand, remain, occur）句型中的 there 已经失去了原有的意义，翻译时可以省译。如：

With the explosively growing reliance on e-mail, there grows a demand for authentication and confidentiality services.

【译文】由于人们日益依赖电子邮件，所以对身份验证和保密服务要求也越来越高。

引导词 it 不仅可以作形式主语、形式宾语，也可以用于强调句型。如：

It is important to remember that the hydrogen bond is an intermolecular force.

【译文】重要的是要记住氢键是存在于分子间的一种力。（形式主语）

It is only in the last one hundred years that we have determined that chemical elements themselves are composed of proton, neutron, and electrons.

【译文】在最近的 100 年里，我们才确定化学元素本身是由质子、中子和电子组成的。（强调句）

二、重复内容的省译

（一）重复出现词语的省译

1. 省译代词。

主要有人称代词、物主代词、指示代词、关系代词四种代词的省译。如：

All you have to do is look at a calculator, a digital clock or watch, or a laptop computer and you see liquid crystals.

【译文】只要看一下计算器、电子钟、电子表或笔记本计算机，就可以看懂液晶。（第二人称代词作主语）

在并列和复合句中，相同的人称代词作主语时，一般只译出前一个即可。如：

Another benefit of fiber is believed to be its ability to lower blood cholesterol.

【译文】据说，纤维的另一个益处是能降低血液中的胆固醇。（省译物主代词 its）

2. 省译名词。

重复的名词在一个句子中再次出现时不需翻译。另外，科技英语中，有时一个概念用两个意义相同或相近的词表达，这时只需翻译其中一个即可。如：

A complex password is a password that has at least eight characters and consists of upper case and lowercase letters, as well as numbers or other symbols.

【译文】一个复杂的密码应至少有八个字符，包括大小写字母、数字和其他符号。

3. 省译动词。

在主从复合句中，主从句中的谓语动词如果表达同一意义，则可以省译。如：

An IGCC plant emits fewer pollutants into the air than conventional coal-fired plants do.

【译文】整体煤气化联合循环发电厂所排放的污染物要比传统的燃煤电厂少。

4. 省译形容词。

Smart or intelligent materials have the ability to adapt to their surroundings.

【译文】智能材料能适应周围环境的变化。

（二）修辞性省译

英汉语言在修辞上存在较大差异，英语比较注重语言形式，是显性的；而汉语更关注语言意义，是隐性的。所以，翻译时可以将"显而易见"的内容删除不译，达到此时"无形胜有形"的修辞效果。如：

To get a good understanding of how this occurs, it's useful to look at a waterfall diagram showing when each resource is downloaded.

【译文】为了更好地理解此过程，可以使用瀑布图显示每个资源的下载过程。（这里前半句已经表达了目的，所以 its useful to 在中文中显得多余，可以省译。）

Section Five　Writing：An Event Invitation Letter

询问出席聚会意愿

一、写作要点 Key Points

Step 1：说明来信的目的是邀请出席聚会。
Step 2：清楚说明聚会时间与地点，并提供相关必要信息。
Step 3：请求对方给予回复。

二、实际 E-mail 范例

| 写信▼ | 删除 | 回复▼ | 寄件者： | Ginny |

Dear Sir/Madam,

Introduction
I am writing this letter on behalf of my company to invite you to our year-end banquet.

Body
The banquet is scheduled to be held on December 23, 7:00 pm—9:00 pm at Regent Hotel. This is a major event of our company and it would be a great honor for us to have you at the banquet.

Closing
Please R. S. V. P. to let us know whether you will be able to attend.
We look forward to seeing you on December 23 at 7 pm at Regent Hotel.

Best regards,
Signature of Sender/Sender's Name Printed

Unit Three　Automation

| E-mail 中译 | 写信▼ | 删除 | 回复▼ | 寄件者： | Ginny |

您好：

开头
　　我谨代表本公司邀请您参加我们的年终宴会。
本体
　　此次宴会预定于12月23日晚上7—9点，于晶华酒店举行。这是本公司的重要活动，您的参与将是我们莫大的荣幸。
结尾
　　请回复此信，好让我们知道您是否可以出席。期待12月23日晚上7点能在晶华酒店见到您。

敬祝安康，
签名档/署名

三、各段落超实用句型

说明：画下划线部分的单词可按照个人情况自行替换。

（一）开头 Introduction

❶On behalf of Mr. Smith, I would like to invite you and your family to the wedding of his daughter, Miss Lauren Smith's wedding party.
我谨代表史密斯先生，邀请您与您家人参加他的千金——史密斯·罗伦小姐的结婚派对。

❷It is with great pleasure that we invite you and your family to our celebration party.
很高兴能邀请您与您的家人来参加我们的庆功宴。

❸In celebration of the launch of our new products, we are giving a cocktail party and we'd like to invite you to join us.
为了庆祝新品上市，我们将举办一场鸡尾酒宴会，并邀请您参加。

❹I am writing this letter to invite all of you to the newcomers' welcome party.
我写这封信是为了邀请各位参加新来同仁的欢迎会。

❺We would like to take the opportunity to invite you to be our guest at our Christmas party.
我们想利用这个机会邀请您担任我们圣诞派对的嘉宾。

（二）本体 Body

❶It would be a great honor for us to have you at the party.
您的参与将是我们莫大的荣幸。

❷The charity dinner will start at 7 pm.
慈善宴会将会在晚上 7 点开始。

❸The party will be held at the Grand Royal Hotel with a black tie dress code.
本次派对将会在皇冠酒店举行，需着正式服装出席。

❹The purpose of the party is to raise funds for charity and education.
本次派对的目的是为慈善及教育机构募款。

❺We have also invited some popular singers to perform at the party for your entertainment.
我们也邀请了一些知名歌手到会场表演助兴。

（三）结尾 Closing

❶We sincerely look forward to your esteemed attendance at the event.
我们诚挚地期望您能纡尊降贵，共襄盛举。

❷Please kindly send us a confirmation reply as soon as possible.
请您尽快给我们确认回复。

❸I look forward to the confirmation of your attendance.
期待能得到您出席的确认信。

❹Look forward to seeing all of you at the welcome party.
期待能在欢迎会上见到各位。

❺Please feel free to contact me if you have any questions or need any additional information.
如果您有任何问题或需要其他信息，请尽快与我联络。

❻If you would like any further information regarding the event such as the dress code and parking availability, please do not hesitate to let me know.
若您需要更多关于此活动的进一步信息，如服装规定及是否方便停车，请不吝告知我。

四、不可不知的实用 E-mail 词汇

newcomer welcome party 迎新会

farewell party 送别会

celebration party 庆功宴

cocktail party 鸡尾酒宴会

retirement party 退休派对

tea party 茶会

year-end party 年终派对

annual banquet 年终宴会

Christmas party 圣诞派对

charity dinner 慈善宴会

Section Six Culture Tips

无人工厂

1984年4月9日，世界上第一座实验用的"无人工厂"在日本筑波科学城建成，并开始进行试运转。这个项目是日本政府通产省工业技术院在20家企业的配合下，从1977年开始筹建的，共耗资137亿日元。

"无人工厂"又叫自动化工厂、全自动化工厂，是指全部生产活动由电子计算机进行控制，生产第一线配有机器人而无须配备工人的工厂。无人工厂里安装有各种能够自动调换的加工工具。从加工部件到装配以致最后一道成品检查，都可在无人的情况下自动完成。

无人工厂的试运转证明，以往需要用近百名熟练工人和电子计算机控制的最新机械，花两周时间制造出来的小型齿转机、柴油机等，只需要用四名工人花一天时间就可以制造出来，这是智能化改造优势的体现。

智能化改造的工业应用是一个繁杂的系统工程，而"无人工厂"的关键性技术则包括柔性化生产技术，工业机器人控制技术，整体安全和监控技术，数据通信技术等。"无人工厂"能使生产率成倍提高，进一步加快了整个制造业自动化的进程。

而近40年后的今天，无人工厂在世界范围内的普及面越来越广，被装上了"智慧大脑"，将人工智能、云计算、大数据、5G、数字孪生等新兴技术广泛应用到工厂产销环节中，成为现代制造业的新生力量。

Section Seven Self-evaluation

Rate your learning outcomes in this unit.

Evaluation Grades		Items				
		A	B	C	D	E
Attitude	I can take the initiative to preview before class.					
	I can take an active part in class activities.					
	I can finish tasks carefully and independently.					
Knowledge	I can make good use of words and expressions concerning automation.					

续表

Evaluation Grades		Items				
		A	B	C	D	E
Ability	I can read the passage about automated factory and automation technologies.					
	I can understand and explain the advantages and disadvantages of automation.					
	I can write an event invitation letter.					
Quality	I have a deep understanding of the importance of automation technologies.					
	I have made up my mind to study my major well.					
Is there any improvement over the last unit?		YES			NO	

千里之行,始于足下。 ——《老子》

A journey of a thousand mile① begins with the first step. —Lao zi

Unit Four

Workplace Safety

Focus

Section One	Warming Up
Section Two	Dialogue: Talking about Electrical Safety
Section Three	Passage: Safety Rules in the Workshop
Section Four	Translation Skills
Section Five	Writing: A Quotation Request Letter
Section Six	Culture Tips
Section Seven	Self-evaluation

---

① 1 mile (mi) = 1.609 km.

Learning Objectives

Upon completion of the unit, students will be able to:

1. Have basic knowledge of safety rules in the workshop.
2. Fully understand the protective and precautionary measures for safety at work.
3. Put "safety first" in mind, enhance safety awareness and follow safety rules at work.

Section One Warming Up

Warming Up 听力

Group Work

Match the pictures with its corresponding meaning.

A. Safety gloves.   B. Safety glasses.   C. Protective mask.
D. Protective shoes.   E. Protective clothing.   F. Helmet.

Listen to the dialogue and find out what Sally advises Yang Kang to do for the new employees.

Unit Four   Workplace Safety

Section Two Dialogue: Talking about Electrical Safety

Listen and role-play the following dialogue.

谈论用电安全

Sally: What's wrong, Yang Kang?
Yang Kang: There's no **power**.
Sally: Have you checked the **fuse** box?
Yang Kang: Yes, the fuse had **blown** and I've changed it, but now the **motor** keeps cutting off.
Sally: There might be a loose connection somewhere that's making the safety **switch trip**. If you can't **fix** it yourself, call in an **electrician**, or it could be dangerous.
Yang Kang: Okay, thank you. It seems that it's necessary for us to know some electricity knowledge in the workshop.
Sally: **Definitely**! Especially the safety rules of electricity must be grasped.
Yang Kang: Could you give me some tips?
Sally: All right. Before you work, make sure you are trained in electrical safety. And you must follow the safety **instructions** at all times.
Yang Kang: I see.
Sally: Then, **identify** all of the possible electric sources that could cause **hazards**. Never **assume** that the equipment or system is **de-energized**. Remember to test it before you touch it. Lock out/tag out machinery or other equipment you will work on. Turn off power before servicing.
Yang Kang: That sounds very useful.
Sally: It is also important to choose the right personal protective equipment (**PPE**).
Yang Kang: I got it. Thank you very much.

Words & Expressions

electrical	[ɪˈlektrɪkəl]	adj.	电动的，电的
power	[ˈpaʊə(r)]	n.	电力（physics）
fuse	[fjuːz]	n.	保险丝，熔丝
blow	[bləʊ]	v.	（保险丝）熔断

motor	[ˈməʊtə(r)]	n. 发动机
switch	[swɪtʃ]	n. 开关
trip	[trɪp]	v.（部分电路）自动断开；跳闸
fix	[fɪks]	vt. 修理；校准
electrician	[ɪˌlekˈtrɪʃn]	n. 电工，电学家
definitely	[ˈdefɪnətli]	adv. 一定地，肯定地
instruction	[ɪnˈstrʌkʃn]	n. 命令，指示
identify	[aɪˈdentɪfaɪ]	vt. 认出，识别
hazard	[ˈhæzəd]	n. 危险
assume	[əˈsjuːm]	vt. 假定，猜想
de-energize	[ˈdəˈenədʒaɪz]	vt. 切断……的电源

fuse box	保险丝盒
cut off	切断
call in	召集；召来
make sure	确保，确信
lock out/tag out	上锁挂牌
turn off	关闭
PPE	个人防护用品

GO Find Information

Task Ⅰ. Read the dialogue carefully and then answer the following questions.

1. What's wrong with the machine?

2. Has Yang Kang checked the fuse box?

3. What does Sally advise Yang Kang to do?

4. Can you list some safety rules of electricity?

5. What is PPE short for?

Task Ⅱ. Read the dialogue carefully and then decide whether the following statements are true (T) or false (F).

(　　) 1. Yang Kang finds something is wrong with the machine.

(　　) 2. Sally advises Yang Kang to fix the machine by himself.

Unit Four　Workplace Safety

(　　) 3. Don't operate machine without being trained in electrical safety.
(　　) 4. Test the machine when it is running.
(　　) 5. Turn off electricity after servicing.

Cheer up Your Ears

DIALOGUE CHEER UP EARS

Listen and write down what you've heard. Then read and recite till you can use them fluently.

1. If you can't fix it yourself, call in an _____, or it could be dangerous.
2. It seems that it's necessary for us to know some _____ knowledge in the workshop.
3. Especially the safety _____ of electricity must be grasped.
4. Before you work, make sure you are trained in electrical _____.
5. You must follow the safety _____ at all times.
6. Then identify all of the possible electric sources that could cause _____.
7. Never assume that the equipment or _____ is de-energized.
8. _____ /tag out machinery or other equipment you will work on.
9. _____ power before servicing.
10. It is also important to choose the right personal _____ equipment.

Words Building

Task Ⅰ. Choose the best answer from the four choices A, B, C and D.

(　　) 1. _____ the switch when anything goes wrong with the machine.
A. Turn up　　B. Turn down　　C. Turn off　　D. Turn on

(　　) 2. He had his finger _____ in an accident at work.
A. cut down　　B. cut off　　C. cut in　　D. cut back

(　　) 3. We need a/an _____ to fix our light switch.
A. politician　　B. magician　　C. musician　　D. electrician

(　　) 4. We had to _____ a plumber to unblock the drain.
A. call off　　B. call in　　C. call up　　D. call by

(　　) 5. Something must _____ to avoid or reduce accidents.
A. be done　　B. do　　C. have done　　D. to be done

Task Ⅱ. Fill in each blank with the proper form of the word given.

1. The _____ industry consumes large amounts of fossil fuels. (electric)
2. The equipment could be _____ if mishandled. (danger)
3. Dust masks are graded according to the _____ they offer. (protect)
4. The new workers need safety _____ for safe production. (train)

5. Remember _____ the power supply before leaving the workshop. (check)

Task Ⅲ. Match the following English terms with the equivalent Chinese items.

A. safety hazard   1. (　) 断路器
B. conduct electricity   2. (　) 触电
C. overvoltage protection 3. (　) 保险丝盒
D. circuit breaker   4. (　) 绝缘鞋
E. fuse box   5. (　) 灭火器
F. insulating gloves   6. (　) 高压电
G. electric shock   7. (　) 安全隐患
H. insulating shoes   8. (　) 过（电）压保护
I. high voltage   9. (　) 绝缘手套
J. fire extinguisher   10. (　) 导电

Table Talk

Pair Work

Role-play a conversation about electric safety with your partner.

Situation

Imagine you are a new employee in the workshop. You want to ask some questions about electric safety to your master. Now you are talking with your partner.

Section Three　Passage：Safety Rules in the Workshop

Pre-reading Questions

1. Have you ever looked through the English job ads?
2. What qualities do you think are necessary for a qualified employee?

车间健康与安全

Attention must be paid to safety in order to ensure a safe working practice in factories. Workers must be aware of the dangers and risks that exist all around them：two out of every three industrial accidents are caused by **individual** carelessness. In order to avoid or reduce accidents, both protective and **precautionary measures** must be followed while working.

1. Always listen carefully to the **production supervisor** and follow instructions. Do not use a machine if the supervisor has not shown you safely.

Unit Four Workplace Safety

2. Report any **damage** to the machines to the supervisor immediately lest they cause an accident.

3. All staff shall wear work clothes and **badge**. In addition, **technicians** and **maintenance** personnel shall wear safety glasses.

4. Always wear a protective **apron** as it will protect your clothes and hold **loose** clothing such as ties in place.

5. Protective shoes must be worn, which can prevent or **minimize** foot injury. No **slippers**, **sandals** or **sneakers** are allowed in the workshop.

6. Always be patient and never **rush** in the workshop.

7. Do not run in the workshop, as you may "**bump**" into a machine or another person and cause an accident.

8. Bags should not be brought into the workshop as people may **trip** over them.

9. Know where the **emergency** stop **buttons** are in the workshop.

10. Do not operate a machine if you feel uncomfortable.

Words & Expressions

individual	[ˌɪndɪˈvɪdʒuəl]	adj. 个别的，个人的
precautionary	[prɪˈkɔːʃənəri]	adj. 预防的
measure	[ˈmeʒə(r)]	n. 措施，办法
production	[prəˈdʌkʃn]	n. 生产，制作
supervisor	[ˈsuːpəvaɪzə(r)]	n. 监督人；主管人
damage	[ˈdæmɪdʒ]	n. 损害，损毁
badge	[bædʒ]	n. 徽章，证章，标志
technician	[tekˈnɪʃn]	n. 技术人员，专家
maintenance	[ˈmeɪntənəns]	n. 维持，维护；保养
apron	[ˈeɪprən]	n. 围裙
loose	[luːs]	adj. 宽松的；未绑紧的
minimize	[ˈmɪnɪmaɪz]	vt. 把…减至最低数量〔程度〕
slipper	[ˈslɪpə(r)]	n. 拖鞋
sandal	[ˈsændl]	n. 凉鞋
sneaker	[ˈsniːkə(r)]	n. 运动鞋
rush	[rʌʃ]	vt. & vi. 冲，奔
bump	[bʌmp]	vt. & vi. 撞倒；冲撞
trip	[trɪp]	vi. & vt. 绊倒
emergency	[iˈmɜːdʒənsi]	n. 紧急情况，不测事件
button	[ˈbʌtn]	n. 按钮

自动控制英语

pay attention to	注意；留意
be aware of	意识到，察觉到
damage to...	对……的损害，损坏
in addition	另外，此外
in place	在适当的地方，在恰当的位置
trip over...	被……绊倒

GO Find Information

Task Ⅰ. Read the passage carefully and then answer the following questions.

1. Why must both protective and precautionary measures be followed while working?
_____

2. Why do you report any damage to the machines to the supervisor immediately?
_____

3. Why must you wear protective shoes?
_____

4. Why can't you run in the workshop?
_____

5. Why should bags not be brought into the workshop?
_____

Task Ⅱ. Read the passage carefully and then decide whether the following statements are true (T) or false (F).

(　　) 1. All of industrial accidents are caused by individual carelessness.
(　　) 2. Always wear your own shirt when using a machine.
(　　) 3. You can wear sneakers in the workshop.
(　　) 4. Damaged tools and machines can be dangerous.
(　　) 5. Keep working when you're sick.

Cheer up Your Ears

PASSAGE CHEER UP EARS

Listen and write down what you've heard. Then read and recite till you can use them fluently.

1. Attention must be paid to _____ in order to ensure a safe working practice in factories.

2. Workers must be aware of the _____ and risks that exist all around them: two out of every three industrial accidents are caused by individual _____.

3. Always listen carefully to the production supervisor and follow _____.

4. Report any _____ to the machines to the supervisor immediately lest they cause an accident.

Unit Four   Workplace Safety

5. In addition, technicians and maintenance personnel shall wear _____.

6. Protective shoes must be worn, which can prevent or minimize _____ injury.

7. Always be _____ and never rush in the workshop.

8. _____ should not be brought into the workshop as people can trip over them.

9. Know where the emergency _____ buttons are in the workshop.

10. Do not operate a machine if you feel _____.

Words Building

Task Ⅰ. Translate the following phrases into English or Chinese.

1. 安全守则   _____

2. 防护围裙   _____

3. 工作服   _____

4. 防护眼镜   _____

5. 电气安全   _____

6. production supervisor   _____

7. industrial accidents   _____

8. personal protective equipment   _____

9. emergency stop button   _____

10. protective and precautionary measures _____

Task Ⅱ. Choose the best answer from the four choices A, B, C and D.

(   ) 1. Wear protective shoes _____ the splinters on the floor should hurt your feet.
A. even though   B. as if   C. lest   D. so that

(   ) 2. Ensure the guard is _____ before operating the machine.
A. in the place   B. in place   C. in place of   D. in the place of

(   ) 3. Don't disturb technician Zhang, _____ is repairing the machine.
A. which   B. who   C. whom   D. that

(   ) 4. We need to be aware _____ the risks and keep working to alleviate the dangers.
A. of   B. on   C. with   D. to

(   ) 5. Some people worried that the rapid development of industry might cause damage _____ the local environment.
A. of   B. on   C. with   D. to

Task Ⅲ. Fill in each blank with the proper form of the word given.

1. Education would be the best and final _____ against social evils. (prevent)

2. The technicians are required to wear protective glasses to offer quite a safe _____. (protect)

3. Open the _____ exit and check the stairs. (emergent)

4. Protective shoes are not as _____ as sneakers. (comfort)

5. Most accidents are caused by ignorance, _____, or lack of skill. (careless)

Task Ⅳ. Complete the sentences below with the correct form of the words and phrases in the box.

| in order to | danger | protective glasses | pay attention to |
| industrial accident | trip over | in addition to | operate |

1. Don't leave tools lying around in the workshop — someone might _____ them.

2. When they set out, they were not aware of the risks and _____ ahead.

3. _____ pollution and traffic jams, auto safety is also a critical issue in developing nations.

4. _____ take that job, you must have left another job.

5. This is an _____ because an employee has been injured by the machine.

6. What skills are needed to _____ this machine?

7. Eventually, the design team equipped themselves with gloves and _____.

8. Safety is the first, which will end your life if you don't _____ it.

Task Ⅴ. Translate the following sentences into Chinese.

1. Workers must be aware of the dangers and risks that exist all around them: two out of every three industrial accidents are caused by individual carelessness.

_____

2. Report any damage to the machines to the supervisor immediately lest they cause an accident.

_____

3. Always wear a protective apron as it will protect your clothes and hold loose clothing such as ties in place.

_____

4. Do not run in the workshop, you may "bump" into a machine or another person and an accident may occur.

_____

5. Know where the emergency stop buttons are in the workshop.

_____

Listening

Task Ⅰ. Listen to 5 short dialogues and choose the best answer.

1. A. Swim in a pool. B. Do exercise in a park.
   C. Walk around a lake. D. Read aloud by a river.

A级听力单选

2. A. At a hospital. B. At an office.
   C. In a classroom. D. In a cafe.
3. A. Fruits. B. Vegetables.
   C. Machines. D. Clothes.
4. A. Selling the car. B. Buying a new car.
   C. Repairing the car. D. Renting a car.
5. A. Make an appointment. B. Go sightseeing.
   C. Book a flight ticket to Beijing. D. Make a dinner reservation.

Task Ⅱ. Listen to the passage and fill in the blanks with the missing words or phrases.

I promise you are going to enjoy your stay here in our city. This is a beautiful, quiet city where you can 1. _____, sit by the beach, enjoy great meals and feel safe. You can walk into town and enjoy the fountains or 2. _____ along the waterside, Please do not swim here. This is not a safe place to swim for its 3. _____ undercurrents. Sanya is the place to go if you want to enjoy swimming 4. _____. You can take a short 5. _____ from your hotel.

 Extensive Reading

A级听力填空

Directions: *Read the text about fire safety procedures and put the actions in the correct order.*

Fire Safety

A fire safety plan is required in all public buildings, from schools, hospitals, supermarkets to workplaces. Generally, the owner of the building is responsible for the preparation of a fire safety plan. Once the plan has been approved by the Chief Fire Official, the owner is responsible for training all staff in their duties.

Evacuation drills are a very important part of the staff training associated with emergency evacuation procedures. Drills should be carried out in all buildings at least once a year. The drill should be checked, recording the time required to complete the evacuation, and without any problems and deficiencies. After each drill, a meeting should be held to evaluate the success of the drill and to solve any problems that may have arisen.

What to do in case of fire?

1. If you see fire or smoke, do not panic. Remain calm and move quickly, but do not run.

2. Alert the responsible staff and dail the correct national emergency number. Have someone meet the firefighters to tell them where the fire is. They can lose valuable minutes if they have to find it themselves.

3. Rescue any people in immediate danger only if it is safe to do so.

4. If practicable, close all doors and windows to contain the fire and prevent it from spreading.

5. Try to extinguish the fire using appropriate firefighting equipment only if you are trained and

it is safe to do so.

6. Follow the instructions of your supervisor and prepare to evacuate if necessary.

7. Save records if possible.

8. Evacuate your area and check all rooms, especially changing rooms, toilets, storage areas, etc.

9. Do a head count of all staff and report any people unaccounted for to the supervisor.

a ☐ Close all doors and windows.
b ☐ Do a head count of all staff and visitors.
c ☐ Evacuate your area and check all rooms.
d ☐ Meet the firefighters and give them details about the fire.
e ☐ Save records.
f ☐ Prepare to evacuate.
g ☐ Remain calm and move quickly.
h ☐ Report any people unaccounted for to the supervisor.
i ☐ Rescue any people in immediate danger.
j ☐ Telephone the correct national emergency number.
k ☐ Try to extinguish the fire using appropriate firefighting equipment.

Section Four　Translation Skills

科技英语翻译技巧——转译法

转译是根据目的语的语言习惯和规律，把源语中的句子成分转换词性后进行翻译。转译包括词类的转译和句子成分的转译。

一、词类的转译

（一）转译成动词

汉语句子大量使用动宾式、连动式和兼语式动词词组或短语，使用的动词较多，有时一个句子甚至存在多个动词，使汉语具有明显的动词化特征，这就导致英译汉时要将英语中的许多词类转译成动词。

1. 英语名词转译成汉语的动词。

表示动作、行为或动作的结果和状态等具有动作意义的普通名词和由动词派生出来的名词，英译汉时，可以转译成动词。如：

Proper selection of circuit components permits a transistor to operate in this characteristic region.

【译文】正确地选择电路元件能使晶体管在这个特殊领域工作。

虚化动词词组中的名词译成汉语动词，英语中有许多虚化动词（light verbs），它们在语义上虚化，但在搭配上却得到强化。虚化动词的词组结构通常为"虚化动词＋名词＋介词"。翻译时，只要将该词组中的名词转译成动词即可。如把 make mention of, take into consideration, keep in mind, make use of 等词组中的 mention, consideration, mind, use 等名词转译成汉语动词。如：

To make a clear interpretation of the UV-VIS spectrum, linear regression analysis was used.

【译文】为了清楚地分析紫外－可见吸收光谱，利用了线性回归分析法。

2. 英语形容词转译成汉语的动词。

英语中一些表示情感、知觉、愿望、态度等心理状态的形容词常跟在系动词后面构成复合谓语，结构"be/get/其他系动词＋形容词＋介词短语或从句"。这些形容词在译成汉语时，需要转性为汉语动词，如：

Online scavenger hunts are easy to create and the resulting interactive searches are both fun and informative for students.

【译文】在网上进行"拾荒式"搜索容易在交互式活动中产生和获得结果，对学生来说既有趣又能增加见闻。

3. 英语介词转译成汉语的动词。

英语中大多数介词含义灵活，一词多义、多用，而汉语介词的数量有限，从而使大多数英语介词不可能对等地翻译成汉语介词，而是转译成汉语的其他词类。多数情况下，英语介词在一些搭配中表达动作意义，可以转译为汉语动词。

英语中的前置介词，在很多情况下，可译成汉语动词或副动词，常见的有对于、关于、因为、为了、根据、按照、把、被、于、在、经过、到、沿、过、给、趁、跟、用、拿、对、比、应、靠、往、去、当、上、下、随等。如：

For a description of standard objects and modules, see the *Python Library Reference Document*.

【译文】需要有关标准对象和模块说明的话，请查询《Python 库参考手册》。

（二）转译成名词

1. 英语形容词转译成汉语名词。

英语中作表语使用表示事物属性特征的形容词，一般都可以转译为汉语的名词。有些形容词在转译成汉语的名词时，可以在其后添加"性、度、体"等缀词。如：

While a high speed charge may be more convenient, a trickle charger ensures longer battery life.

【译文】尽管快速充电会带来更多的方便，但是点滴式充电器却能延长电池的寿命。

2. 英语动词转译成汉语名词。

英语中一些由名词派生而来的动词，其概念很难用汉语动词来表达，翻译时可译成汉语名词。

Those who have computer culture will succeed in the information economy while those who lack it will inevitably fall between the socioeconomic cracks.

【译文】具有计算机文化的人将会在信息经济中取得成功，而缺乏这种文化的人不可避免地落入社会经济学的裂缝中。

3. 英语副词转译成汉语名词。

Carburetors worked fairly well but could be mechanically complex.

【译文】化油器性能良好，但是其机械操作比较复杂。

（三）转译成形容词

1. 由形容词派生来的名词或一些作表语且前面没有定冠词的名词，可以转译成形容词。如：

Many users realize the importance of confidential information when it is stored on their workstations or servers.

【译文】许多用户意识到，当信息存储在工作站或服务器时，信息保密非常重要。

2. 英语副词转译成汉语形容词。

由于某些英语的动词在汉译时转译成汉语的名词，英语中原来修饰动词的副词应随之转译成汉语相应的形容词。如：

The computer is chiefly characterized by its accurate and quick computations.

【译文】这台计算机的主要特点是准确性高且速度快。

（四）形容词转译成副词

由于某些英语的名词在汉译时转译成汉语的动词，所以，英语中原来修饰名词的形容词应随之转译成汉语相应的副词。如：

Fast Ethernet limits the distance between a computer and a hub to only 100 m—making careful network planning a necessity.

【译文】快速以太网把计算机与集线器之间的距离限制为仅100 m，这就必须细心地设计网络。

二、句子成分的转译

英译汉的时候，不仅需要对词类进行转译，有时甚至需要转译句子成分，才能使译文达到逻辑正确、通顺流畅、重点突出，句子成分转译的运用范围也相当广泛，常见的主要有以下几种情况。

（一）英语句子的主语转译成汉语的宾语、定语、状语、表语等

1. 英语的主语转译成汉语的宾语。如：

Some work is still needed to be done on the congestion control of this new system to make it more efficient in any kind of network.

【译文】为了使新的系统在任何一种网络上都行之有效，还需要再做一些工作应对其拥塞控制。

2. 英语的主语转译成汉语的定语。如：

The network manager is responsible for the connection from the PC to the server.

【译文】网络管理器的任务是将计算机和服务器连接起来。

3. 英语的主语转译成汉语的状语。如：

Any virus signatures stored in the database must be carefully handled.

【译文】务必对存储在数据库中的病毒特征码进行妥善处理。

4. 英语的主语转译成汉语的表语。如：

A typical example of remote forwarding is the following.

【译文】下面是远程转发的一个典型例子。

（二）英语句子的宾语（表语）转译成汉语的主语

英语句子的宾语或者表语可以转译成汉语的主语，如：

Revenue generation should play a more important role in the internet of things, to generate new money streams.

【译文】为了获取新的资金流，创收在物联网中的角色显得越来越重要。

（三）英语句子的状语转译成汉语的主语

英语句子中的状语可以转译成汉语的主语，如：

The most stunning transitions were introduced in PowerPoint 2010 and sadly aren't available in PowerPoint 2007.

【译文】PowerPoint 2010 版推出了许多非常好的初换效果，很可惜这些在 2007 版上没有。

Section Five　Writing：A Quotation Request Letter

要求厂商报价

一、写作要点 Key Points

Step 1：请求对方对某服务或产品提供报价。

Step 2：请求报价，或提供对方报价所需相关信息。

Step 3：请对方在具体时间内回复报价。

二、实际 E-mail 范例

| 写信▼ | 删除 | 回复▼ | 寄件者： | Ginny |

Dear Sir/Madam,

Introduction
I am writing this letter for a price quote on replacing the central air conditioning system in our office.

Body
The footage of our office is approximately 6,840 m^2 with three individual central air conditioning units. It is our hope that the work can be done in less than one week, so please take that into account in your price quote.

Closing
Please provide us with pricing information and the estimated length of work you need to complete the job.
We look forward to hearing from you as soon as possible.

Best regards,
Signature of Sender/Sender's Name Printed

E-mail 中译 | 写信▼ | 删除 | 回复▼ | 寄件者： | Ginny |

您好：

开头
　　谨以此信请您提供更换本公司办公室中央空调系统的报价。
本体
　　我们办公室的面积大约是 6 840 m^2，并且有三个独立中央空调机。我们希望能在一周内完工，所以报价时请把这点考虑进去。
结尾
　　请给我们提供报价，以及您估计完工所需的时间。
　　我们期待能尽快得到您的回复。

敬祝安康，
签名档/署名

Unit Four　Workplace Safety

三、各段落超实用句型

说明：画下划线部分的单词可按照个人情况自行替换。

（一）开头 Introduction

❶We are interested in purchasing your <u>Color Laserjet Pro MFP 3300</u> for our office use.
我们有兴趣采购贵公司生产的<u>彩色激光专业级打印机，型号为 MFP3300</u>，作为办公用途。

❷I am writing this letter to inquire about the price of your <u>multi-function printer MFP3300</u>.
我写这封信是为了询问贵公司<u>多功能打印机 MFP3300</u> 的价格。

❸We are currently planning to refurbish the office of our <u>Hong Kong Branch</u>.
我们目前计划要翻新我们<u>香港分公司</u>的办公室。

❹We would like to request a quotation on the following items.
我们想要请你们提供下列产品的报价。

❺We would be very grateful to you if you could provide us with the <u>quotation of your lift repair and maintenance service</u>.
如果您能提供<u>电梯维修服务</u>的<u>报价</u>给我们，我们将非常感激。

❻I'm writing to ask for a quotation on the <u>office file cabinets</u> your company provides.
我写这封信是希望能够得到您公司的<u>办公室文件柜</u>产品的报价。

（二）本体 Body

❶We are in urgent need of <u>1,000 multi-function printers</u>.
我们急需<u>一千台多功能打印机</u>。

❷The items we need are listed as below.
我们所需要的物品详列如下。

❸We would like to request you to send us your <u>quotation</u> at the earliest.
我们想您尽快将<u>报价单</u>寄给我们。

❹Please include <u>packing and delivery</u> in your quoted prices.
请将<u>包装及运送</u>费用包含在您的报价中。

❺Please send us your <u>sales quotation with terms of payment</u>.
请将<u>付款条件连同产品报价</u>一同寄给我们。

❻We would highly appreciate a list of the <u>prices for each item</u> as well as the available <u>payment methods</u>.
若能获得<u>各产品的价格清单</u>及可用的<u>付款方式</u>，我们将感激不尽。

（三）结尾 Closing

❶Your prompt reply would be highly appreciated.
若您能尽快回复，我们将不胜感激。

❷We would appreciate your quotation by <u>the end of the week</u> for your consideration.

希望您能在这周结束之前提供报价单让我们考虑，感激不尽。

❸We hope to receive your quotation no later than this Friday.
我们希望能在本周五之前收到您的报价单。

❹Please send us your quotation by November 18，2023 for our acquisition consideration.
请在2023年11月18日之前将您的报价单寄给我们，以便我们考虑采购。

❺Provided the prices meet with our budget, we would place our order within a week.
如果价格符合我们的预算，我们将会在一周内下单。

❻As the renovation process will be starting soon, we would be very grateful if we could receive your quotation before the end of this week.
由于要开始重新装潢了，我们非常希望能够在这周结束前获得您的报价。

四、不可不知的实用 E-mail 词汇

request for quotation 请求报价
provide the quotation 提供报价
meet with designated budget 符合既定预算
acquisition consideration 采购考虑
service charge 服务费
packing charge 包装费
delivery charge 运费

Section Six　Culture Tips

新《安全生产法》简介

　　2021年6月10日，第十三届全国人民代表大会常务委员会第二十九次会议通过了关于修改《中华人民共和国安全生产法》（以下简称《安全生产法》）的决定，自2021年9月1日起施行。现行《安全生产法》于2002年制定，经过2009年和2014年两次修改，现在是第三次修改，这部法律对预防和减少生产安全事故发挥了重要作用。

　　但是，过去长期积累的传统隐患还没有完全消除，有的还在集中暴露，新的风险又不断涌现，虽然全国生产安全事故总体上呈下降趋势，但目前又开始进入一个瓶颈期，而且稍有不慎，重特大事故还会反弹。

　　同时新发展阶段、新发展理念、新发展格局又对安全生产工作提出了更高的要求，因此，安全生产工作仍处于爬坡、过坎期。

Unit Four Workplace Safety

在这个,阶段尤其是全国开展安全生产三年行动、制定实施"十四五"安全生产规划的关键时期,对安全生产法进行修改,正当其时、十分必要,为安全生产工作提供了有力的法律武器。

这次修改《安全生产法》,贯彻了新发展理念,贯彻落实了党中央关于安全生产的有关重大决策部署,始终坚持问题导向,认真总结实践经验,积极回应社会关切问题。这次修改力度大,共修改决定42条,大约占原来条款的1/3,主要包括以下五方面的内容。

一是完善工作原则要求。

二是完善安全监管体制。

三是强化企业主体责任。

四是强化政府监督管理职责。

五是加大违法处罚力度。

Section Seven Self-evaluation

Rate your learning outcomes in this unit.

Evaluation Grades		Items				
		A	B	C	D	E
Attitude	I can take the initiative to preview before class.					
	I can take an active part in class activities.					
	I can finish tasks carefully and independently.					
Knowledge	I can make good use of words and expressions concerning workplace safety.					
Ability	I can read the passage about electric safety and safety rules in the workplace.					
	I can understand, explain and follow safety rules in the workplace.					
	I can write a quotation request letter.					
Quality	I have the consciousness of safety production.					
	I will work safely.					
Is there any improvement over the last unit?		YES			NO	

满招损，谦受益。 ——《尚书》
Complacency leads to failure; modesty to success. —The Book of Rites

Unit Five

Autonomous Cars

Focus

Section One	Warming Up
Section Two	Dialogue: How Far Away Are We from Autonomous Cars?
Section Three	Passage: Google's Self-driving Vehicles
Section Four	Translation Skills
Section Five	Writing: A Shipment Request Letter
Section Six	Culture Tips
Section Seven	Self-evaluation

Unit Five Autonomous Cars

Learning Objectives

Upon completion of the unit, students will be able to:

1. Have basic knowledge of autonomous cars.
2. Fully understand the technology used in autonomous cars.
3. Keep in mind the advantages of autonomous cars.

Warming Up 听力

Section One Warming Up

Group Work

Do you know the following famous autonomous cars in the history?

A. 1925 American Wonder.   B. 1939 Futurama.   C. 2009 Google Driverless Car.

Listen to the dialogue and find out which car they are talking about.

Section Two   Dialogue: How Far Away Are We from Autonomous Cars?

Listen and role-play the following dialogue.

Sally: Nowadays, **automatic** driving has become one of the selling points of cars, but in fact, it is not a new concept, and it has a long history of nearly a hundred years.

Yang Kang: Really? Then when did people begin to study **autonomous** cars?

Sally: In August 1925, a wireless remote control vehicle named American Wonder was **officially unveiled**. Francis P. Houdina, an electric engineer of the United States Army, controlled the steering wheel, clutch, brake and other **components** remotely by means of wireless remote control.

Yang Kang: It's hard to imagine. But is it a real autonomous car?

Sally: No. In 1939, General Motors (GM) displayed Futurama, the world's first autonomous concept car. This is an electric vehicle guided by radio controlled **electromagnetic** fields generated by **magnetized** metal spikes **embedded** in the road. However, it was not until 1958 that GM realized this concept.

Yang Kang: Then, what about Google's self-driving vehicle? Is it very famous?

Sally: Since 2009, Google has been secretly developing driverless car projects. In 2014, Google **demonstrated** the prototype of driverless cars without steering wheel, accelerator or brake pedal.

Yang Kang: There is no doubt that the future of autonomous vehicles is **promising**, but in the process of its development, there are also technical and **ethical issues**. The high-level autonomous driving we expect will take time to develop and mature.

Words & Expressions

automatic	[ˌɔːtəˈmætɪk]	adj.	自动的
autonomous	[ɔːˈtɒnəməs]	adj.	自主的
officially	[əˈfɪʃəli]	adv.	官方地
unveil	[ˌʌnˈveɪl]	vt.	揭开, 揭幕

自动驾驶离我们有多远?

Unit Five Autonomous Cars

component	[kəmˈpɒnənt]	n. 零件
electromagnetic	[iˌlektrəʊmæɡˈnetik]	adj. 电磁性的
magnetize	[ˈmæɡnɪˌtaɪz]	v. 使磁化
embed	[emˈbed]	vt. 使……嵌入
demonstrate	[ˈdemənˌstreɪt]	vt. 表明
promising	[ˈprɒmɪsɪŋ]	adj. 有前途的
ethical	[ˈeθɪkəl]	adj. 伦理的
issue	[ˈɪʃuː]	n. 议题

be guided by... 被……指导
no doubt 毫无疑问
be generated by... 由……产生
by means of... 凭借……，借助……

GO Find Information

Task Ⅰ. Read the dialogue carefully and then answer the following questions.

1. How long is the history of automatic driving?

2. When did people begin to study automatic driving?

3. What's the name of the wireless remote control vehicle?

4. Who invented the wireless remote control vehicle?

5. Which company displayed Futurama?

Task Ⅱ. Read the dialogue carefully and then decide whether the following statements are true (T) or false (F).

(   ) 1. Automatic driving is a new concept.

(   ) 2. Francis P. Houdina is an electric engineer of the United States Army.

(   ) 3. American Wonder is the world's first autonomous concept car.

(   ) 4. Since 2014, Google has been secretly developing driverless car projects.

(   ) 5. In 2014, Google demonstrated the prototype of driverless cars without steering wheel, accelerator or brake pedal.

Cheer up Your Ears

Listen and write down what you've heard. Then read and recite till you can use them fluently.

1. Nowadays, _____ driving has become one of the selling points of cars.

2. It has a long _____ of nearly a hundred years.

3. In August 1925, a _____ remote control vehicle named American Wonder was officially unveiled.

4. Francis P. Houdina is an electric _____ of the United States Army.

5. In 1939, General Motors _____ Futurama, the world's first autonomous concept car.

6. This is an _____ vehicle guided by radio controlled electromagnetic fields.

7. Since 2009, Google has been secretly _____ driverless car projects.

8. In 2014, Google _____ the prototype of driverless cars without steering wheel, accelerator or brake pedal.

9. There is no doubt that the future of autonomous vehicles is _____.

10. The high-level autonomous driving we expect will take time to develop and _____.

Words Building

DIALOGUE CHEER UP EARS

Task Ⅰ. Choose the best answer from the four choices A, B, C and D.

(   ) 1. Automatic driving has become one of the _____ of cars.
A. sell point   B. selling points   C. sold points   D. selling point

(   ) 2. We express our thought _____ words.
A. in the means of   B. by mean of   C. in means of   D. by means of

(   ) 3. Automatic driving has a long history of nearly _____.
A. a hundred years   B. a hundred year   C. hundred years   D. hundreds years

(   ) 4. _____ she'll call us when she gets there.
A. No doubts   B. Not doubt   C. No doubt   D. No doubts

(   ) 5. Most electricity is _____ by steam turbines.
A. to generate   B. generating   C. generate   D. generated

Task Ⅱ. Fill in each blank with the proper form of the word given.

1. The women who believe in Islam often mask their faces with _____ before going out. (unveil)

2. The company's _____ retiring age is 65. (officially)

3. The loop becomes _____ when the current is switched on. (magnetize)

4. _____ has helped to increase production. (automatic)

5. I thought it was not _____. (ethics)

Task Ⅲ. Match the following English terms with the equivalent Chinese items.

A. meter　　　　　　　　1. (　) 方向盘
B. steering wheel　　　　2. (　) 主驾座椅
C. driving seat　　　　　3. (　) 安全带
D. seat belt　　　　　　　4. (　) 备胎
E. parking lever　　　　　5. (　) 里程表
F. jack　　　　　　　　　6. (　) 启动
G. spare tyre　　　　　　7. (　) 中排气管
H. fuel tank　　　　　　　8. (　) 手刹
I. middle pipe　　　　　　9. (　) 千斤顶
J. starter　　　　　　　　10. (　) 油箱

 **Table Talk**

Pair Work

Role-play a conversation about autonomous cars with your partner.

Situation

Imagine you are a freshman in the company and want to know about autonomous cars. You are talking with the senior employee, Sally on the subject.

Section Three　Passage: Google's Self-driving Vehicles

Pre-reading Questions

1. Do you know anything about Google's self-driving vehicles?
2. How many components are mentioned in the passage?
3. Do you want to have an autonomous car in the future?

谷歌无人驾驶汽车

Google's self-driving vehicles understand where they are and what's around them through **sensors** that are purpose-built to help the vehicles **perceive** their **surroundings** accurately, and software that processes the information received.

1. Laser sensor.

This sensor gives the vehicle a 360° understanding of its **environment** so the vehicle can sense objects in front of, beside, and behind itself at the same time. The laser also helps the vehicle to **determine** its **location** in the world.

2. Safety drivers.

Drivers also test the vehicles daily, reporting feedback on how to make the ride more safe and comfortable.

3. Processor.

Information from the sensors is cross-checked and processed by the software so that different objects around the vehicle can be sensed and **differentiated** accurately, and safe driving decisions can then be made based on all the information received.

4. Position sensor.

This sensor, located in the wheel hub, detects the **rotations** made by the wheels of the car to help the vehicle understand its position in the world.

5. **Orientation** sensor.

Similar to the way a person's **inner** ears, it gives them a sense of motion and balance, this sensor, located in the **interior** of the car, works to give the car a clear sense of orientation.

6. Radar.

This sensor detects vehicles far ahead and measures their speed so that the car can safely slow down or speeds up with other vehicles on the road.

Words & Expressions

sensor	[ˈsensə]	n. 传感器
perceive	[pəˈsiːv]	vt. 意识到
surrounding	[səˈraʊndɪŋ]	n. 周围的事物，周围环境
environment	[enˈvairənmənt]	n. 环境，周围的状况或条件
determine	[dɪˈtɜːmin]	vt. 判定
location	[ləʊˈkeiʃən]	n. 地点
differentiate	[ˌdifəˈrenʃiˌeit]	vt. 区分，在两事物间区分差别
rotation	[rəʊˈteiʃən]	n. 旋转，转动
orientation	[ˌɔːriːenˈteiʃən]	n. 趋向
inner	[ˈinə]	adj. 内部的，发生在或位于内部的
interior	[inˈtiəriə]	adj. 内部的

at the same time	同时
be based on...	基于……
be located in	位于
be similar to...	与……相似
slow down	减速
speed up	加速

GO Find Information

PASSAGE CHEER UP EARS

Task Ⅰ. Read the passage carefully and then answer the following questions.

1. How many components are mentioned in the passage?
_____

2. What do safety drivers do?
_____

3. What can radar do?
_____

4. Which component is similar to a person's inner ears?
_____

5. Where is the position sensor?
_____

Task Ⅱ. Read the passage carefully and then decide whether the following statements are true (T) or false (F).

(　　) 1. Google's self-driving vehicles understand where they are and what's around them.

(　　) 2. Safety drivers give the vehicle a 360° understanding of its environment.

(　　) 3. The laser also helps the vehicle to determine its location in the world.

(　　) 4. Information from the sensors is cross-checked and processed by the software.

(　　) 5. Radar is similar to the way a person's inner ears that give them a sense of motion and balance.

Cheer up Your Ears

Listen and write down what you've heard. Then read and recite till you can use them fluently.

1. _____ gives the vehicle a 360° understanding of its environment.

2. The laser also helps the vehicle to _____ its location in the world.

3. Google's self-driving vehicles understand where they are and what's around them through _____ that are purpose-built to help the vehicles perceive their surroundings accurately, and

— 87 —

_____ that processes the information received.

4. Drivers also test the vehicles daily, reporting _____ on how to make the ride more safe and comfortable.

5. Information from the sensors is cross-checked and _____ by the software so that different objects around the vehicle can be sensed and differentiated accurately.

6. Safe driving _____ can then be made based on all the information received.

7. Position sensor, located in the wheel hub, detects the _____ made by the wheels of the car.

8. It helps the vehicle understand its _____ in the world.

9. Orientation sensor is similar to the way a person's _____ ears that give them a sense of motion and balance.

10. This sensor, located in the interior of the car, works to give the car a clear _____ of orientation.

Words Building

Task Ⅰ. Translate the following phrases into English or Chinese.

1. 感知环境 _____
2. 处理信息 _____
3. 决定位置 _____
4. 报告反馈 _____
5. 安全舒适 _____
6. position sensor _____
7. orientation sensor _____
8. inner ear _____
9. a sense of motion _____
10. a sense of balance _____

Task Ⅱ. Choose the best answer from the four choices A, B, C and D.

(   ) 1. The two runners broke the tape _____.
A. at the same time   B. at same time   C. in the same time   D. in same time

(   ) 2. It costs a lot of money if it's made to _____ the original.
A. be similar to   B. be similar in   C. be similar at   D. be similar on

(   ) 3. The novel was based _____ real life.
A. with   B. in   C. on   D. to

(   ) 4. What _____ it is!
A. a useful information   B. an useful information
C. useful information   D. useful an information

(   ) 5. Different objects around the vehicle can be sensed and differentiated _____, and _____ driving decisions can then be made based on all the information received.

A. accurately; safe B. accurately; safely

C. accurate; safe D. accurate; safely

Task Ⅲ. Fill in each blank with the proper form of the word given.

1. How do you account for the _____ between them? (differentiate)

2. They decided to _____ a new school in the suburbs. (location)

3. The following action steps are a part of the foreman's _____ of the new worker. (orientate)

4. Our sales people need _____ and drive. (determine)

5. The company's headquarter _____ in London. (base)

Task Ⅳ. Complete the sentences below with the correct form of the words and phrases in the box.

be based on	location	slow down	speed up
surround	in	differentiate	safe

1. Animals in zoos are not in their natural _____.

2. The TV play _____ real life.

3. _____ your car is important before you turn.

4. My husband has several shirts of _____ colors.

5. It's very important to teach the children about road _____.

6. The holidays simply _____.

7. This city _____ on the southern fringe of the desert.

8. She went into the _____ room to change her dress.

Task Ⅴ. Translate the following sentences into Chinese.

1. Information from the sensors is cross-checked and processed by the software so that different objects around the vehicle can be sensed and differentiated accurately.

2. This sensor, located in the wheel hub, detects the rotations made by the wheels of the car to help the vehicle understand its position in the world.

3. The car can sense objects in front of, beside, and behind itself at the same time.

4. Orientation sensor is similar to the way a person's inner ear gives them a sense of motion and balance.

5. This sensor, located in the interior of the car, works to give the car a clear sense of orientation.

 **Listening**

A 级听力填空

Task Ⅰ. Listen to 5 short dialogues and choose the best answer.

1. A. Applying for a visa.　　　　　　B. Making a reservation.
   C. Checking out in the airport.　　　D. Filling in an application form.
2. A. In an office.　　　　　　　　　B. In a club.
   C. In a hospital.　　　　　　　　　D. In a restaurant.
3. A. From a TV ad.　　　　　　　　B. From a friend.
   C. From a newspaper.　　　　　　　D. From the radio.
4. A. Check the statistics.　　　　　　B. Visit some clients.
   C. Draw some charts.　　　　　　　D. Conduct a survey.
5. A. Mr. Brown isn't fit for the job.　　B. Mr. Brown can do the job well.
   C. She doesn't know Mr. Brown.　　D. She can do the job herself.

A 级听力填空

Task Ⅱ. Listen to the passage and fill in the blanks with the missing words or phrases.

Good evening, ladies and gentlemen! On behalf of our company, I'd like to thank you for coming to 1. _____ the opening of our new branch office in Hattiesburg. This branch is the 10th office we have 2. _____ in the country. I'm glad we finally opened a branch in the southeast area. Now, I would like to 3. _____ to thank all the staff here for your efforts to establish the branch. In order to successfully operate the branch, we need the 4. _____ of customers like you being present. We will do our best to provide you with the 5. _____ . Thank you very much.

 Extensive Reading

Directions: *After reading the passage, you are required to choose the best answer from the four choices for each statement.*

Autonomous Cars

The first truly autonomous cars—vehicles that drive down the streets with no one behind the wheel—have finally arrived.

Waymo, which began its life as Google's self-driving car project, announced that it had let its driverless cars free in parts of Phoenix, Arizona, with nobody in the front seats to take over in case of emergency. Members of the public taking part in a Waymo trial in the desert city in the southwest U.S. will be able to order the vehicles to come through an app "in the next few months", the group said.

Unit Five Autonomous Cars

Possibly one of the most revolutionary new technologies, as well as one of the most advertised, driverless cars have been at the centre of a race between big automakers and technology companies. But while a number of groups are testing the technology on the streets with back-up drivers behind the wheel, most believe the approach of full autonomy is at least two years away. Google shocked the auto industry when it first showed a basic version of its driverless technology seven years ago, and it then invested more than $1bn in autonomous vehicle research.

Rivals admit it still has a technology lead, though skeptics question whether the artificial intelligence is good enough to respond to many unforeseen events that could occur on the road.

Waymo believes it is the first company to reach a standard known in the driverless car world as level 4, meaning its cars can drive under full autonomy in preset areas that have been carefully mapped out and tested. Uber, General Motors, Aptiv PLC, BMW and others have been conducting testing to reach level 4, but all of them still keep a human in the driver's seat.

1. What should people do in the driverless car test?

A. Take over in an emergency.

B. Call for a car without an app.

C. Let the car operate in full autonomy.

D. Test the car on a land.

2. The fourth paragraph may be used to show _____.

A. Google's driverless technology is really advanced

B. Waymo will be widely acknowledged two years later

C. driverless cars will damage the entire automobile market

D. the technology of driverless cars is developing too fast

3. Why is Waymo confident about trying out its first driverless cars?

A. Because it already invented a basic version.

B. Because it has a technology lead in this area.

C. Because it has reached full, level 4 autonomy.

D. Because its cars can drive wherever a person likes.

4. Which statement is FALSE according to the passage?

A. The first truly autonomous cars have finally arrived.

B. A number of groups are testing the technology on the streets without back-up drivers behind the wheel.

C. Google's driverless technology is one of the most revolutionary new technologies.

D. Google shocked the auto industry when it first showed a basic version of its driverless technology.

5. What's the best title of the text?

A. New Technologies   B. The Artificial Intelligence

C. A Technology Lead   D. The Self-driving Cars

Section Four　Translation Skills

科技英语翻译技巧——换序法

换序法是指将英语与句子中的词、词组或者从句的位置进行变动，或置于句首，或置于句尾，或插入句中。有时候还需要对原文中多个部分的位置进行调换和重新组合。

一、定语的换序

（一）定语单词换序翻译

英语中有时将单词置于名词后面作后置定语，翻译成汉语时，需要符合汉语的表达习惯，进行定语前置。

1. 当后置定语是以-able/-ible 词缀结尾的形容词，且所修饰的中心语前面又有 only、every、all，或形容词最高级以及其他限定词时，这些后置定语应进行换序翻译。如：

Oxygen contains only eight electrons available, two in the first shell and six in its outer shell.

【译文】氧原子仅有八个电子，其中第一层两个，最外层六个。

2. 当具有较强动词含义的现分在词或过去分词充当后置定语时，需要将其前置。如：

Distillation can remove the methanol produced.

【译文】蒸馏法可以去除所产生的甲醇。

3. 英语中修饰 something、anything 或 nothing 等不定代词的定语通常放在不定代词的后面，在汉译时需要前置。如：

Chemical changes result in the production of something different from the original substances.

【译文】化学变化会产生和原物质不同的新物质。

4. 作表语的形容词通常充当后置定语，如 present、akin、alike、alone、alive、available 等，汉译时，需要前置。如：

There are over fifty thousand links to sports sites alone from this one website.

【译文】在这一个网站里，仅仅体育方面的网址链接就超过 5 万多条。

5. 副词 here、there、up、down、above、below 等充当后置定语修饰名词，汉译时需前置。如：

The accident review above has shown that the most frequent fatal and serious injury is a frontal collision.

【译文】上面的事故评估报告显示大多数致命和重伤事故都是由正面碰撞造成的。

（二）定语词组换序翻译

1. 介词短语换序翻译。如：

Everything around us is made up of elements.

【译文】我们周围的所有物质都是由元素组成的。

2. 形容词短语换序翻译。科技英语中部分形容词具有动词的含义，而且这类形容词多以-able或-ible作为后缀，如obtainable、suitable和responsible等。它们修饰中心语时往往置于所修饰的中心词之后，汉译时需要将它们前置。如：

Theoretically there would be no limit to the power obtainable from the engine.

【译文】理论上讲，可从发动机上获得的能量将会是无限的。

3. 分词短语换序翻译。英语中现在分词和过去分词与所修饰的中心语构成逻辑上的主谓关系，汉译时要前置。如：

Almost all interactions between machines are based upon the protocols outlined above.

【译文】机器之间几乎所有的相互交流都是基于上述协议进行的。

4. 不定式短语换序翻译。不定式短语常用于修饰中心语，并与其构成逻辑上的主谓结构，汉译时也需要将其前置。如：

The user notification setup screen displays the administrators to be notified when viruses are detected.

【译文】当检测到病毒时，"用户通知设置"屏幕即显示需要通知的管理员名单。

（三）定语从句换序翻译

定语从句通常放在所修饰的中心语之后。限制性定语从句与前面的先行词关系较紧密，汉译时，需要将定语从句前置；而非限定性定语从句与前面的修饰语关系较弱，主要是补充新信息。如：

The computer you use will also have to have iTunes, which is a software program you download and install from Apple's offical website.

【译文】你使用的计算机需安装有iTunes播放软件程序，它可以从苹果官网上下载。

二、状语的换序翻译

（一）状语单词换序翻译

英语单词作状语时，可以放在动词之前或之后，而汉语中的状语则放在谓语动词之前。所以，后置状语单词汉译时需要前置。如：

The time and date for your system should be set when you first turn on your computer.

【译文】第一次打开计算机时，应该设置系统的时间和日期。

（二）状语短语换序翻译

1. 表示原因的状语往往由because of、due to、on account of、as a result of等引导，汉译时需要放在句首。如：

Capacitors have the ability to pass AC current because of their constant charging and discharging action.

【译文】电容器由于在不断地充电和放电，所以能通交流电。

2. 表目的的状语往往由 for 等介词引导，在汉译时要对置于句尾的目的状语换序，将其放在句首，如：

Contents filtering is a necessary component for e-mail security.

【译文】为了邮件安全，需要过滤邮件内容。

3. 表示条件的状语可以由 if、after、without 等词引导，汉译时要注意换序。如：

Whichever program you use, you should back up your data at least weekly—if not daily.

【译文】不管你使用什么程序，如果不是每天备份数据，至少应每周一次。

4. 由 in spite of 和 despite 等词引导的让步状语可以置于英语句尾，但汉译时往往放在句首。如：

Online shopping isn't a security risk, in spite of the concerns of many net newbies.

【译文】尽管许多网络新手会有所顾虑，但是网上购物并不存在安全风险。

5. 英语中的方式状语常常跟在动词后面，但是在汉译时，需要将它们放在所修饰的动词前面。如：

The driver can choose to change gears in manual mode by moving the gearshift.

【译文】司机可以通过移动变速杆选择手动换挡。

6. 在英语中，时间状语可以放在句子的开头或结尾，但在汉语中，时间状语则应放在句首或主语后面，所以汉译时要将置于句尾的时间状语前置。如：

Windows XP will not have the same memory constraints that many other Windows versions ran into after running for a few hours.

【译文】在计算机工作了几个小时之后，Windows XP 不会像许多其他版本 Windows 操作系统那样出现内存不足的现象。

7. 在英语中，地点状语可以放在句尾，汉译时，要视汉语句子的表达习惯或放在句首，或放在句中。如：

You can customize the way the buttons and bars appear on your browser, adding things that are missing or removing things you don't use.

【译文】你可以定制按钮和工具栏在浏览器中出现的方式，如添加新的按钮和工具栏，或者删除不使用的按钮和工具栏。

（三）状语从句换序翻译

英语句法一般遵循"主句在前，从句在后"的规律。汉语则是先偏后正。因此，英译汉时需适时调整语序。如：

The webpage will not change its own after you place the page on your desktop.

【译文】当你将网页保存在桌面上，网页上的内容自己不会发生改变。

Unit Five　Autonomous Cars

三、其他结构的换序翻译

（一）同位语的换序

在汉译时，同位语往往置于中心语之前，作定语处理，也可以放在句首，单独译成句子。如：

The fact that the newly created element is unstable and may last for only a fraction of a second before it decays makes detection difficult.

【译文】产生的新元素往往不稳定，只能存在一小会儿，这给识别工作带来了难度。

（二）插入语的换序

英语中存在诸多插入语，如 of course、however、for example、in total、in contrast 等。它们在句子中的位置相当灵活，汉译时，可以将它们置于句首。如：

In France, for example, the acceptance of nuclear power is much higher than in other countries.

【译文】譬如说，法国民众对核能的接受程度要比其他国家高很多。

（三）倒装句的换序

为了强调某部分内容，会使用倒装句。在翻译成汉语时，在不影响原文内容的前提下，可以将倒装句换成陈述句。如：

Only in extreme circumstances should knowledgeable user cancel a running process.

【译文】计算机高手只有在万不得已的情况下才会取消程序运行。

四、it 引导的主语从句的换序

it 引导的主语从句结构复杂，而且句子较长，需将这些从句单独翻译，并置于句首。如：

Although it's probably impossible to do away with 100% of the spams you receive, there are steps you can take to reduce the amount of spams you have to deal with.

【译文】尽管不太可能杜绝垃圾邮件的骚扰，但是可以采取一些方法来减少你需要处理的数量。

Section Five　Writing：A Shipment Request Letter

催促出货

一、写作要点 Key Points

Step 1：说明来信的目的与某订单有关。

Step 2：提醒对方应交货日期，并表示对方已延迟出货。
Step 3：催促对方立刻出货，并告知对方对延迟出货的处理方式。

二、实际 E-mail 范例

| 写信▼ | 删除 | 回复▼ | 寄件者： | Ginny |

Dear Sir/Madam,

Introduction
I am writing this letter with regard to the order for 20 milemeters that we placed with you on April 30.
Body
According to our agreement, our order should be shipped within five working days once our payment is settled. That is to say, we should have received our order by last Friday. However, our commodities haven't been delivered to us so far.
Closing
Should you fail to deliver our order by the end of this week, we will have no option but to exercise our right to terminate our agreement with you and seek compensation from your company.

Best regards,
Signature of Sender/Sender's Name Printed

| E-mail 中译 | 写信▼ | 删除 | 回复▼ | 寄件者： | Ginny |

您好：

开头
　　我写的这封信是关于我们在 4 月 30 日向您订购 20 台里程表的订单。
本体
　　根据我们的合作协议，我们订购的商品应该在账款支付之后五个工作日内出货，也就是说，我们在上周五之前就应该要收到我们订购的物品，然而我们订购的商品一直到现在还没有送达。
结尾
　　如果您未能在本周结束前帮我们出货，我们将不得不行使我们的权利终止与您的合作，并根据协议向贵公司要求赔偿。

敬祝安康，
签名档/署名

Unit Five　Autonomous Cars

二、各段落超实用句型

说明：画下划线部分的单词可按照个人情况自行替换。

（一）开头 Introduction

❶Our order is supposed to be delivered to us two weeks ago.

我们订购的商品在两周前就应该寄给我们了。

❷I am writing this letter to inform you of the delay of delivery of 500 umbrellas that we ordered from you.

我写这封信是来通知您，我们向贵公司订购的500 把雨伞交货延迟了。

❸We regret to point out that we have not yet received our order, which should be shipped to us seven days ago.

我们必须很遗憾地告诉您，我们还没有收到七天前就应该送来的订单货品。

❹You accepted our order that we placed on September 15 and claimed to ship our order to us by September 22.

您受理了我们9 月15 日下的订单，并声称会在9 月22 日前出货。

❺I'm writing this letter to let you know that our order has not yet reached us.

写这封信是要让您知道，我们订购的商品到现在还没送来。

（二）本体 Body

❶It should have arrived two days ago.

两天前就该到了。

❷According to our contract, our order should be shipped within two weeks after the order was accepted.

根据我们的合约，我们的订单应该在受理订单后两周内出货。

❸It has been three weeks since the payment of my order was settled in full, however, I still haven't received my commodities.

我的订单账款三周前就已全额结清，然而我还没有收到商品。

❹You agreed to have our order shipped to us by September 30, but we have not received it up to this date.

您同意在9 月30 日前将我们的订单出货，但我们至今仍未收到物品。

❺I was surprised to find that my order hasn't arrived yet, even though you agreed to ship it a month ago.

我很惊讶地发现，虽然您同意一个月前就要送出我的订货，我却现在都还没收到。

（三）结尾 Closing

❶Thank you for your prompt attention to this matter.

感谢您立即处理这个问题。

❷We must ask you to have our order shipped to us as soon as possible.

我们必须要求贵公司立即将我方订购的货物出货。

❸If you cannot proceed with the shipment immediately, we have to cancel our order.
若您无法立刻出货，我们将不得不取消订单。

❹The delay of the delivery has been causing considerable inconvenience to our company.
延迟出货为我们公司带来相当大的困扰。

❺Please understand that this delay has caused us great loss in business.
请您明白这次延迟出货已经造成我们业务上极大的损失。

❻We would terminate our contract with you if this happens again.
若这种事情再次发生，我们将终止与贵公司的合约。

四、不可不知的实用 E-mail 词汇

delay of delivery 延迟出货
loss in business 业务上的损失
terminate the contract 终止合约
cancel the order 取消订单
seek compensation 寻求赔偿

Section Six　Culture Tips

自动驾驶重塑未来

　　1939 年，通用汽车公司在纽约世博会上展出了自动驾驶概念车 Futurama。此后，许多科技发达国家都开展了自动驾驶技术的初步探索和基础研究。作为人工智能具有商业价值的领域之一，自动驾驶产业广受热捧，发展迅速。国际电气与电子工程师协会预测，到 2040 年全球 75% 的新款汽车都将会是自动驾驶的。自动驾驶在推动人类交通革命的同时，也必将塑造城市的未来。

　　目前，世界各国正在积极开展自动驾驶技术的基础研究，抢占人工智能的竞争高地。2016 年，日本发布《自动驾驶普及路线图》，允许自动驾驶汽车 2020 年在高速公路上通行。2017 年，德国通过了针对自动驾驶汽车的法案，并于同年提出了自动驾驶指导原则。我国在 2017 年发布实施《新一代人工智能发展规划》和《国家车联网产业标准体系建设指南（智能交通相关）》。当前，我国在自动驾驶道路诊断方面取得了突破性进展。2017 年 12 月，深圳较早地在开放道路上进行了无人驾驶公交的试运行诊断。此后，上海、北京、重庆相继为自动驾驶诊断车辆颁发了诊断牌照。根据麦肯锡的研究，中国未来可能成为全球较大的自动驾驶市场。

　　然而，目前国际上尚无统一的自动驾驶等级划分标准。我国正在积极制定自动驾驶等级划分标准。美国曾将自动驾驶技术分为五个等级。国际汽车工程师学会按照智能化程度将自动驾驶汽车划分为完全人类驾驶、辅助驾驶、部分自动驾驶、有条件自动驾驶、高度自动驾

驶、完全自动驾驶六个标准等级。据比标准，当前配备自动驾驶功能的汽车几乎都在四级标准以下，且处于基本诊断阶段。2017年，加利福尼亚常规公路诊断里程数较高的自动驾驶汽车厂商，诊断里程总数不超过40万千米，技术人员无干预的行驶里程也低于6 000千米。此外，路测也仅于乘用车场景，并没有实现多个行业都已经出现的应用场景的全覆盖，距离真正意义上完全自动驾驶的目标还有相当长的距离。

Section Seven Self-evaluation

Rate your learning outcomes in this unit.

Evaluation Grades		Items				
		A	B	C	D	E
Attitude	I can take the initiative to preview before class.					
	I can take an active part in class activities.					
	I can finish tasks carefully and independently.					
Knowledge	I can make good use of words and expressions concerning autonomous cars.					
Ability	I can understand history and introduction of autonomous cars.					
	I can tell the main components of autonomous cars.					
	I can write a proper shipment request letter.					
Quality	I can understand the advantages and disadvantages of autonomous cars.					
	I can tell my idea about autonomous cars.					
Is there any improvement over the last unit?		YES		NO		

穷则变，变则通，通则久。　　　　　　　　——《易经》
Extreme—Change—Continuity　　　　　—The Book of Changes

Unit Six

High Speed Trains

Focus

Section One	Warming Up
Section Two	Dialogue: History of High Speed Trains
Section Three	Passage: High Speed Trains in China
Section Four	Translation Skills
Section Five	Writing: A Late Delivery Apology Letter
Section Six	Culture Tips
Section Seven	Self-evaluation

Learning Objectives

Upon completion of the unit, students will be able to:
1. Have basic knowledge of high speed trains.
2. Fully understand the history and technology of high speed trains.

Unit Six High Speed Trains

3. Keep in mind the advantages of high speed trains.

Section One Warming Up

Warming Up 听力

Group Work

Do you know the following famous high speed trains in history?

A. Shinkansen.   B. ETR 200 train.   C. Harmony.

_____

_____

_____

Listen to the dialogue and find out when the first high speed train appeared in China.

Section Two Dialogue: History of High Speed Trains

Listen and role-play the following dialogue.

高铁的历史

Yang Kang: What are you doing, Sally?

Sally: I am reading a book about high speed trains.

Yang Kang: What kind of locomotives can be called the high speed trains?

Sally: Well, there are different **standards** of what **constitutes** high speed trains based on the train's speed and **technology** used. In the European Union, high speed trains are those travel 125 mi/h or faster, while in the U. S. they are those that travel 90 mi/h or faster.

Yang Kang: When and where did the first high speed trains **appear**?

Sally: The first high speed trains appeared as early as 1933 in Europe and the U. S.

Yang Kang: Is Shinkansen a kind of high speed trains?

Sally: Yes, in the mid-1960 s, Japan **introduced** the world's first high **volume** high speed train that **operated** with a standard (4 ft①) **gauge**. It was called Shinkansen and officially opened in 1964. It **provided** rail service between Tokyo and Osaka at speeds of around 135 mi/h. These are the early **development** of the high speed trains.

Yang Kang: Then, do you know when the high speed trains first appeared in China?

Sally: In 2005, Beijing-Tianjin **Intercity** Railway.

Words & Expressions

standard	[ˈstændəd]	n.	标准
constitute	[ˈkɔnstitjuːt]	vt.	构成,构成……部分或成分;组成
technology	[tekˈnɔlədʒi]	n.	技术
appear	[əˈpiə]	vi.	出现
introduce	[ˌintrəˈdjuːs]	vt.	提出
volume	[ˈvɔljuːm]	n.	容积
operate	[ˈɔpəˌreit]	vt.	操作:执行一项功能,工作
gauge	[geidʒ]	n.	轨距:铁路两根钢轨之间的距离
provide	[prəˈvaid]	vi.	供应,提供

① 1ft = 0.3048 m.

| development | [dɪˈveləpmənt] | n. 发展，开发 |
| intercity | [ˌɪntəˈsɪti] | adj. 城市间的 |

as early as... 早在……时
in the mid-1960s 在20世纪60年代
at speeds of... 以……的速度

GO Find Information

DIALOGUE CHEER UP EARS

Task Ⅰ. Read the dialogue carefully and then answer the following questions.

1. What kind of trains can be called high speed trains in the European Union?
_____

2. What kind of trains can be called high speed trains in the United States?
_____

3. When and where did the first high speed trains appear?
_____

4. Which country introduced the world's first high volume high speed train that operated with a standard gauge?
_____

5. When did the high speed trains first appear in China? And what's the name of it?
_____

Task Ⅱ. Read the dialogue carefully and then decide whether the following statements are true (T) or false (F).

(　　) 1. Yang Kang is reading a book about high speed trains.
(　　) 2. There is a same standard of what constitutes high speed trains based on the train's speed and technology used.
(　　) 3. In the European Union, high speed trains are those which travel 125 mi/h or faster.
(　　) 4. The first high speed trains appeared as early as 1933 in Europe and the U. S.
(　　) 5. The high speed trains first appeared in China in 2005.

Cheer up Your Ears

Listen and write down what you've heard. Then read and recite till you can use them fluently.

1. Then, do you know when the high speed trains first _____ in China?
2. It _____ rail service between Tokyo and Osaka at speeds of around 135 mi/h.
3. These are the early _____ of the high speed trains.
4. Japan _____ the world's first high volume high speed train that operated with a standard gauge.

5. While in the United States, they are those that travel 90 mi/h or _____.

6. In the European Union, high speed trains are those which _____ 125 mi/h or faster.

7. There are different standards of what constitutes high speed trains based on the train's speed and _____ used.

8. What kind of locomotives can be _____ the high speed trains?

9. I am reading a book about _____.

10. Beijing-Tianjin _____ Railway was the first high speed trains that appeared in China?

Words Building

Task Ⅰ. Choose the best answer from the four choices A, B, C and D.

(　　) 1. He gave us a brief _____ about his company.
A. introduction　　　B. introducing　　　C. introductions　　　D. introduced

(　　) 2. The article is full of _____ terms.
A. technique　　　B. technical　　　C. technology　　　D. technician

(　　) 3. Germany is _____ European country.
A. an　　　B. the　　　C. /　　　D. a

(　　) 4. China's oil industry is a _____ industry.
A. development　　　B. developed　　　C. developing　　　D. develop

(　　) 5. In the European Union, high speed trains are those _____ travel 125 mi/h or faster.
A. which　　　B. whose　　　C. who　　　D. what

Task Ⅱ. Fill in each blank with the proper form of the word given.

1. The government saw the _____ of new technology as vital. (introduce)

2. The dolphin has evolved a highly _____ jaw. (develop)

3. The _____ of Internet marked the beginning of the age of information. (appear)

4. The hotel _____ a reservation of tickets for its residents. (provide)

5. The country's _____ embodies the ideals of freedom and equality. (constitute)

Task Ⅲ. Match the following English terms with the equivalent Chinese items.

A. business class　　　　　　1. (　　) 测试阶段
B. speed cut　　　　　　　　2. (　　) 防护栏
C. run chart　　　　　　　　3. (　　) 直达列车
D. non-stop train　　　　　　4. (　　) 运行图
E. protect fence　　　　　　　5. (　　) 减速
F. test period　　　　　　　　6. (　　) 商务舱
G. bullet train　　　　　　　　7. (　　) 动车
H. on-schedule rate　　　　　8. (　　) 普通座
I. passage flow　　　　　　　9. (　　) 客流量

J. standard seat 10. (   ) 正点率

 Table Talk

Pair Work

Role-play a conversation about high speed trains with your partner.

Situation

Imagine you want to know more about the high speed trains especially the development of them in China.

Section Three Passage: High Speed Trains in China

Pre-reading Questions

1. What do you know about Chinese high speed trains?
2. Which country is a leader in rail technology?
3. Which country holds the crown of high speed train capital of the world?

It was Japan that was once famous for its high speed train network, introducing the world to its Shinkansen, or **bullet** trains, way back in 1964. Japan is still a leader in rail technology, while it is now China that holds the **crown** of high speed train **capital** of the world.

In twenty years or so since China put into **operation** its first high speed passenger trains, the country has **constructed** more than 40,000 km of high speed rail **track** to create the longest network on Earth. In 2017, the country **launched** the world's fastest high speed train, known as Fuxing, which travels at up to 350 km/h, **reducing** travel time between Beijing and Shanghai to four and a half hours.

Now, China's high speed trains—officially **defined** as passenger trains that travel at speeds of 250 – 350 km/h— take travelers to almost all of the country's provinces. With Inner Mongolia's first high speed line opening in July 2017, only Xizang and Ningxia currently lack high speed trains. But with plans for the continued **expansion** of the network it won't be long until they too are serviced by high speed lines.

All this makes for a super-fast, and relatively inexpensive way, to cover this country's **vast distances**.

Words & Expressions

bullet	[ˈbʊlɪt]	n.	在形状、作用或效果上和射弹相似的东西
crown	[kraʊn]	n.	荣誉：因有功绩而得到的荣誉或奖章
capital	[ˈkæpɪtl]	n.	资源或优势
operation	[ˌɔpəˈreɪʃən]	n.	有效率的经营方式
construct	[kənˈstrʌkt]	vt.	建造；构筑
track	[træk]	n.	轨道；跑道
launch	[lɔːntʃ]	vt.	发起；推出（新产品）
reduce	[rɪˈdjuːs]	vt.	减少；缩小
define	[dɪˈfaɪn]	vt.	为……下定义
expansion	[ɪkˈspænʃn]	n.	膨胀；扩展；扩充
vast	[vɑːst]	adj.	巨大的；广阔的
distance	[ˈdɪstəns]	n.	距离；路程；远方

be known as...	作为……而著名
be famous for...	因……而著名
way back	很久以前
be a leader in...	在……方面领先
be serviced by...	由……提供服务
more than	超过

 Find Information

Task Ⅰ. Read the passage carefully and then answer the following questions.

1. When did Japan introduce its high speed train Shinkansen?

2. Which country is a leader in rail technology?

3. Which country holds the crown of high speed train capital of the world?

4. How long does Fuxing take to travel between Beijing and Shanghai?

5. Till July 2017, which parts of China currently lack high speed trains?

Unit Six　High Speed Trains

Task Ⅱ. Read the passage carefully and then decide whether the following statements are true (T) or false (F).

(　　) 1. It was Japan that once was famous for its high speed train network, introducing the world to its Shinkansen, or bullet trains, way back in 1974.

(　　) 2. Japan is still a leader in rail technology, while it is now China that holds the crown of high speed train capital of the world.

(　　) 3. In twenty years or so since China put into operation its first high speed passenger trains, the country has constructed more than 40,000 km of high speed rail track to create the longest network on Earth.

(　　) 4. Traveling by high speed train is more expensive.

(　　) 5. Inner Mongolia's first high speed line was opened in July, 2017.

Cheer up Your Ears

Listen and write down what you've heard. Then read and recite till you can use them fluently.

1. All this makes for a super-fast, and relatively inexpensive way, to cover this country's _____ .

2. With plans for the continued expansion of the network it won't be long until they too are _____ by high speed lines.

3. With Inner Mongolia's first high speed line opening in July 2017, only Xizang and Ningxia currently _____ high speed trains.

4. Now, China's high speed trains—officially _____ as passenger trains that travel at speeds of 250 – 350 km/h.

5. China's high speed trains take travelers to almost all of the country's _____ .

6. Fuxing travels at up to 350 km/h, _____ travel time between Beijing and Shanghai to four and a half hours.

7. In 2017, China _____ the world's fastest high speed train, known as Fuxing.

8. In twenty years or so since China put into _____ its first high speed passenger trains.

9. The country has _____ more than 40,000 km of high speed rail track to create the longest network on Earth.

10. Japan is still a leader in rail _____ , while it is now China that holds the crown of high speed train capital of the world.

Words Building

PASSAGE CHEER UP EARS

Task Ⅰ. Translate the following phrases into English or Chinese.

1. 减少旅行时间_____

107

2. 因……而著名 ＿＿＿＿＿＿＿＿＿＿＿＿＿＿＿

3. 作为……而著名 ＿＿＿＿＿＿＿＿＿＿＿＿＿＿＿

4. 在……方面领先 ＿＿＿＿＿＿＿＿＿＿＿＿＿＿＿

5. 由……提供服务 ＿＿＿＿＿＿＿＿＿＿＿＿＿＿＿

6. vast distances ＿＿＿＿＿＿＿＿＿＿＿＿＿＿＿

7. relatively inexpensive ＿＿＿＿＿＿＿＿＿＿＿＿＿＿＿

8. continued expansion ＿＿＿＿＿＿＿＿＿＿＿＿＿＿＿

9. rail technology ＿＿＿＿＿＿＿＿＿＿＿＿＿＿＿

10. put into operation ＿＿＿＿＿＿＿＿＿＿＿＿＿＿＿

Task Ⅱ. Choose the best answer from the four choices A, B, C and D.

(   ) 1. She wants to be famous ＿＿＿＿ a movie star.

A. for   B. as   C. to   D. with

(   ) 2. Since China put into operation its first high speed passenger trains, the country ＿＿＿＿ the longest rail tracks.

A. have constructed   B. constructed   C. has constructed   D. are constructing

(   ) 3. It is necessary that he ＿＿＿＿ there at once.

A. will be sent   B. sent   C. will send   D. be sent

(   ) 4. It ＿＿＿＿ be long until the whole country is serviced by high speed lines soon.

A. won't   B. didn't   C. doesn't   D. shan't

(   ) 5. Fuxing travels at up to 350 km/h, ＿＿＿＿ travel time between Beijing and Shanghai to four and a half hours.

A. having been reducing   B. having reduced

C. though reduced   D. reducing

Task Ⅲ. Fill in each blank with the proper form of the word given.

1. The flight was delayed owing to ＿＿＿＿ reasons. (technology)

2. To give a ＿＿＿＿ of a word is more difficult than to give an illustration of its use. (define)

3. Metals undergo ＿＿＿＿ when heated. (expand)

4. The ＿＿＿＿ of this machine is simple. (operate)

5. We sent a lot of people there to help in the ＿＿＿＿. (construct)

Task Ⅳ. Complete the sentences below with the correct form of the words and phrases in the box.

| a leader in | be famous for | be famous as | put into operation |
| at speeds of | way back | make for | the crown of |

1. Honesty keeps ＿＿＿＿ the causeway.

2. Our community is ＿＿＿＿ the conservation of wildlife.

3. Cleanliness ＿＿＿＿ good health.

4. Spain used to ＿＿＿＿ its strong armada.

5. I first met him _____ in the 1980s.

6. Thomas Edison _____ an inventor.

7. Thomas Edison _____ his inventions.

8. The plan began to _____ .

Task Ⅴ. Translate the following sentences into Chinese.

1. Japan is still a leader in rail technology, while it is now China that holds the crown of high speed train capital of the world.

_____

2. In twenty years or so since China put into operation its first high speed passenger trains, the country has constructed more than 40,000 km of high speed rail track to create the longest network on Earth.

_____

3. In 2017, the country launched the world's fastest high speed train, known as Fuxing, which travels at up to 350 km/h.

_____

4. China's high speed trains—officially defined as passenger trains that travel at speeds of 250 – 350 km/h.

_____

5. The development of high speed trains makes for a super-fast, and relatively inexpensive way, to cover this country's vast distances.

Listening

Task Ⅰ. Listen to 5 short dialogues and choose the best answer.

A级听力单选

1. A. A tour to China.　　　　　　B. A business meeting.

　C. A Chinese product.　　　　　D. A job interview.

2. A. The ID card.　　　　　　　　B. The label.

　C. The receipt.　　　　　　　　　D. The menu.

3. A. Buy another smartphone.　　B. Return to the office.

　C. Go to the man's home.　　　D. Cancel their trip.

4. A. Write some invitation letters.　B. Put up a conference poster.

　C. Prepare some documents.　　D. Work out a meeting schedule.

5. A. She volunteered in a museum.　B. She stayed with her parents.

　C. She made a survey in a school.　D. She went traveling abroad.

Task Ⅱ. Listen to the passage and fill in the blanks with the missing words or phrases.

ABC Travel Agency organized a 10-day tour for us to many famous places of interest in China in October last year. They arranged for internal travel by 1. _____ , booked hotels and various guided activities. But we arranged our own 2. _____ to and from China and extensions to the tour to Singapore. ABC Travel Agency was good value for money when 3. _____ other travel agencies. It was about 40% less than I was quoted by well-known UK travel companies for the same itinerary. I would have no hesitation recommending it. Its guides were 4. _____ and generally knowledgeable. Most of them spoke good English. Some even went beyond the agreed itinerary and arranged 5. _____ activities for us.

 Extensive Reading

听力填空

Directions: *After reading the passage, you are required to choose the best answer from the four choices for each statement.*

Nine Fastest Trains in the World

Of course, trains can't fly over oceans like airplanes. But that doesn't mean trains can't run as fast as planes. Fortunately, some trains in this modern world are as fast as planes.

1. Shanghai Maglev, 267.8 mi/h, China.

Shanghai Maglev is the fastest train in the world with a maximum operating speed of 267.8 mi/h. Maglev is an abbreviation of magnetic levitation—suspension or floating of an object by the magnetic field.

2. Harmony CRH 380A, 236.12 mi/h, China.

China railways Harmony CRH 380A is the second fastest operating train service in the world.

3. AVG Italo, 223.6 mi/h, Italy.

AGV Italo is the fastest running train in Europe. The AVG Italo operates between Rome and Naples.

4. Siemens Velaro E/AVS 103, 217.4 mi/h, Spain.

Velaro E12 is the Spanish version of Velaro high speed trains developed by German engineering company Siemens. In Spain, Velaro trains are named as AVS 103. The Velaro E operates between Barcelona and Madrid.

5. Talgo 350, 217.4 mi/h, Spain.

Spain's Talgo 350 high speed train is operated by the state-run railway company. This high speed train can achieve a maximum speed of 217.4 mi/h. The Talgo 350 runs between Madrid and Barcelona.

6. E5 Series Shinkansen Hayabusa, 198.8 mi/h, Japan.

E5 series Shinkansen Hayabusa is the fastest high speed train service in Japan today. This train service started on March 5, 2011 and operated by the East Japan Railway Company.

7. Alstom Euroduplex, 198.8 mi/h, France.

Euroduplex trains are the third series of TGV Duplex high speed trains. The French railway company SNCF operates the Euroduplex train service. This service connects French, Swiss, German and Luxembourg rail network.

8. SNCF TGV Duplex, 198.8 mi/h, France.

TGV5 Duplex is the fastest train service in France. The SNCF6 railway company operates this high speed train service. TGV duplex started the service in December 2011.

9. ETR 500 Frecciarossa, 186.4 mi/h, Italy.

ETR1 500 Frecciarossa 2 is the fastest train service in Italy, operated by Trenitalia company. This service runs between Milan to Naples route and offers 72 connections daily.

1. How many high speed trains are mentioned in the passage? _____.
    A. Ten   B. Eight   C. Nine   D. Seven

2. How many countries are there mentioned in the passage? _____.
    A. Five   B. Six   C. Seven   D. Eight

3. Which of the following train service started on March 5, 2011 and operated by the East Japan Railway Company? _____.
    A. Talgo 350, 217.4 mi/h, Spain
    B. E5 Series Shinkansen Hayabusa, 198.8 mi/h, Japan
    C. ETR 500 Frecciarossa, 186.4 mi/h, Italy
    D. Siemens Velaro E/AVS 103, 217.4 mi/h, Spain

4. Which country has the fastest train in the world? _____.
    A. Italy   B. Japan   C. France   D. China

5. Which service connects French, Swiss, German and Luxembourg rail network? _____.
    A. Alstom Euroduplex, 198.8 mi/h, France
    B. Shanghai Maglev, 267.8 mi/h, China
    C. SNCF TGV Duplex, 198.8 mi/h, France
    D. Siemens Velaro E/AVS 103, 217.4 mi/h, Spain

Section Four　Translation Skills

科技英语翻译技巧——拆译法

拆译法是根据英汉句子的语法结构和表达习惯，在充分理解原文的基础上，将原句中不容易理解的成分分离出来，独立成句，达到化繁为简、化整为零的效果。

一、主语的拆译

英语句子中作主语的可以是单词、词组、从句，有时出现多个词、短语或从句，甚至还夹带定语。对于作主语的单词，照译即可；若复杂的短语和从句作主语，需要将其和谓语拆开处理，可以拆成一个独立的分句，也可以拆成一个独立的简单句。

（一）短语作主语拆译

关键要看拆译成汉语后所形成的独立分句之间的关系是否明确，若逻辑关系已经非常明确，无须增加任何连词；反之，需要增加适当的连词。如：

1. 需要增加连词。

Roentgen's discovery of X-rays stimulated great interest in this new form of radiation worldwide.

【译文】伦琴发现了X射线后，世界范围内掀起了这股新的放射研究热潮。

2. 不需要增加连词。

若拆分后各分句之间的关系比较明确，无须增加任何连词，但有时可以增加一个指示代词"这"，使译文意义更明确。如：

The implementation of these policies by the Financial Services Authority resulted in changes being made in certain consumer protection requirements.

【译文】金融服务管理局实施了这些政策，这使得某些消费者保护要求发生了变化。

英语中，同位语作主语时，如果同位语本身是一个结构复杂且又长的短语，要先将其拆译成一个句子，然后再翻译后面的句子成分。如：

Now the most common means of voice communication in the world, the telephone of today is infinitely more sophisticated and effective than the crude instrument developed by Bell.

【译文】今天的电话已成为当今世界上最普通的通话工具，它比当年贝尔所发明的粗陋的电话机要复杂有效得多。

（二）从句作主语拆译

英语中的主语从句类型多样，如果it所指代的真正的主语结构复杂且句子又长，汉译时需要拆译成一个分句。从句作主语拆译主要有两种方法：一种是按照原文的顺序先翻译前面的部分，再翻译真正主语的部分；另一种是先翻译真正主语的部分，并置于句首，然后翻译剩余的部分，也就是先使用"拆译法"，再使用"换序法"。如：

It is clear that the continued development of more energy efficient technologies will be necessary.

【译文】很显然，人类一直在研发效能更强的技术。

二、谓语的拆译

英语句子中，有时一个谓语动词后面跟着多个宾语（或表语），如果直接翻译，谓语动词有时和其中一个宾语搭配不当。因此，需要将一个谓语动词拆成两个相同或相近的动词，分别搭配其中一个宾语。另外，科技英语的被动结构，有时直接翻译后，译文显得生硬、不自然，所以也要将它们拆分后再翻译。

（一）主动结构的谓语拆译

当动词后面有多个宾语或表语时，可以将谓语动词拆译，并根据宾语或表语的意义选词翻译。如：

Semiconductor materials are neither conductors nor insulators.

【译文】半导体材料既不是导体，也不是绝缘体。

（二）被动结构的谓语拆译

在科技英语的被动结构中，有时会有两个或两个以上的施动者，除了使用换序法和拆译法翻译外，还可以将谓语动词拆开翻译，以对应不同的施动者。如：

The prices of oil are not only affected by supply and demand but also by politics.

【译文】石油价格不仅受到供求关系的影响，还受政治动荡的影响。

三、宾语的拆译

对于简单的宾语，可直接照译；对复杂的宾语，可以采取拆译法，将它们拆译成单独的一个句子。另外，英语中有许多宾语从句，汉译时，可以将它们拆译成单独的句子，使译文更加清晰。如：

Early models of atoms showed electrons orbiting around the nucleus, in analogy with planets around the sun.

【译文】早期的原子模型证明，电子绕着原子核的转动，如同行星绕着太阳转动一样。

四、定语的拆译

（一）短语作定语拆译

当后置定语复杂，或者与所修饰的中心语之间逻辑上存在主谓关系时，可以拆开翻译。如：

A new kind of computer—cheap, small, light, is attracting increasing attention.

【译文】一种新型计算机越来越引起人们的注意，它造价低、体积小、质量轻。

（二）从句作定语拆译

除了可以使用换序法翻译定语从句外，还可以使用拆译法翻译。拆分法主要有下面三种情形。

1. 定语从句与其他成分有明显的逻辑关系，如表示原因、结果、目的等关系，这时只需增加一些连接词或指示代词单独译出即可。如：

This (the user log off function on the start button) is a useful feature in offices where a PC is shared among a number of people.

【译文】因为办公室里一台计算机有多个人共用，所以开始菜单里的注销功能选项很实用。

2. 非限制性定语从句与主句的关系相对松散，可以拆分，单独翻译。如：

Clicking on the word "people" opens a window into which you enter an e-mail address.

【译文】用鼠标单击"people"一词就打开一个窗口，你可以在里面输入电子邮箱地址。

3. 修饰主语的定语从句，在逻辑上所修饰主语构成完整的主谓结构，可以直接拆分翻译。如：

自动控制英语

Chemical cells that may be reactivated by charging are called secondary cells, or storage cells.
【译文】有些化学电池可以通过充电而重复使用，这样的电池叫作二次电池或蓄电池。

五、状语的拆译

状语可以由单个副词或表示时间、地点、方式等的短语或从句来充当，为了使译文结构清晰、语义明了，可以拆译。

（一）单词作状语拆译

有时逐字照译副词，译文生硬晦涩。这时可以单独翻译，保证译文流畅自然。如：
Simply point to the selected menu with the mouse to display details.
【译文】用鼠标单击选中的菜单呈现具体内容，就这么简单。

（二）短语作状语拆译

英语中，介词短语、现在分词或过去分词短语、不定式短语等都可以作状语。当这些短语较长，且与主句的关系相对松散时，可以拆分翻译。如：
In pushing a weight forward we have to exert a force enough to overcome the resistance against the weight.
【译文】只有施加一个足以克服重物阻力的力，才能推动重物。

（三）从句作状语拆译

英语中存在各种类型的从句，可以表示原因、假设、让步、时间等关系。在汉译时，可以将部分从句拆开处理，单独译成一个分句。当然，在拆译时，还需要采用增译法，增加一些指示代词和连词，以使得译文句子通顺。下面的例子就是表示原因的从句拆译：
A computer keyboard has more keys than a typewriter keyboard because it has a number of keys that perform special functions.
【译文】计算机键盘的按键要比打字机键盘多，这是由于前者的许多按键都有专门的功能。

Section Five　Writing: A Late Delivery Apology Letter

延迟出货致歉信

一、写作要点 Key Points

Step 1：为延迟出货道歉。
Step 2：简短说明延迟出货原因，并表示已设法解决。
Step 3：承诺尽快出货，并恳请对方谅解。

二、实际 E-mail 范例

| 写信▼ | 删除 | 回复▼ | 寄件者： | Ginny |

Dear Sir/Madam,

Introduction
With reference to your order for 1,000 protect fences that you placed with us on June 10, we regret to have delayed the delivery.

Body
An unexpected problem occurred in our delivery process, but we managed to fix it this morning and shipped your order immediately. Your commodities should arrive at your appointed address this afternoon.

Closing
Please accept our sincere apology for the inconvenience that this delay has caused you. We assure you that this kind of mistake will not happen again in our future cooperation.

Best regards,
Signature of Sender/Sender's Name Printed

E-mail 中译 | 写信▼ | 删除 | 回复▼ | 寄件者： | Ginny |

您好：

开头
　　有关您在6月10日向我们下的1 000防护栏订单，我们很抱歉延迟了交货时间。
本体
　　我们的出货流程突然发生了一点问题，但是我们已经在今天上午设法解决，并立刻为您出货。您的商品应该在今天下午就会送达您指定的地址。
结尾
　　对于这次延迟交货为您带来的不便，请接受我们的诚挚歉意。我们向您保证，在我们未来的合作中将不会再发生相同的问题。

敬祝安康，
签名档/署名

三、各段落超实用句型

说明：<u>画下划线部分</u>的单词可按照个人情况自行替换。

（一）开头 Introduction

❶We are regretful for delaying your order.
我们对延误您的订单感到很抱歉。
❷With this letter, we would like to express our regret for the late delivery of your order.
我们希望借此信向您表达延迟出货的歉意。
❸We regret to inform you that we delayed your delivery.
很抱歉通知您，我们延迟了交货时间。
❹This letter is to apologize for delaying your shipment.
这封信的目的是为延迟出货而向您致歉。
❺On behalf of my company, I must apologize for the delay in delivering your order.
我谨代表本公司，为延迟寄送您的订单商品而致歉。

（二）本体 Body

❶Our supplier was unable to provide us with sufficient <u>computer parts</u> that we need.
我们的供应商无法提供给我们足够的<u>计算机零件</u>。
❷An unexpected problem occurred in our <u>delivery</u> process, but we managed to fix it <u>this morning</u> and shipped your order immediately.
我们的<u>出货</u>流程突然发生了一点问题，但我们已经在<u>今天上午</u>设法解决，并立刻为您出货了。
❸<u>Typhoon Marge</u> has seriously disrupted our <u>production</u> schedule.
<u>台风"玛琪"</u>严重地打乱了我们的<u>生产</u>进度。
❹The unusual large number of orders during <u>the Chinese New Year Holidays</u> disordered our <u>delivery</u> schedule.
<u>春节假期</u>的庞大订单，打乱了我们的<u>配送</u>时间表。
❺We will ship your order by fast courier immediately. You should be able to receive your commodities within the next <u>two days</u>.
我们将立刻以快递出货给您。您应该在<u>两天</u>内就可以收到商品了。
❻We have adopted a new delivery system to improve our delivery process. Therefore, this kind of delay shall never happen again.
我们已经采用新的出货系统来改善我们的出货流程。因此这样的延迟出货状况未来将不会再发生。
❼<u>A company emergency</u> has prevented us from completing the order as scheduled.
由于<u>公司发生紧急事件</u>，我们无法如期完成订单。

（三）结尾 Closing

❶We apologize for all the trouble that the late delivery has caused.

Unit Six　High Speed Trains

我们为延迟交货所造成的一切困扰致歉。

❷Please accept our sincere apology for the inconvenience that this delay has caused you.
对于这一延误给您带来的不便，请接受我们最诚挚的歉意。

❸We appreciate your patience and your kind understanding.
我们感谢您的耐心及善意体谅。

❹We assure you that this kind of mistake will not happen again in our future cooperation.
我们向您保证，在我们未来的合作中，类似的问题将不会再发生。

❺We are deeply regretful to have caused you such inconvenience, and want to assure you that something like this will never happen again.
给您造成不便，我们非常遗憾，希望您能明白这样的事绝对不会再发生了。

四、不可不知的实用 E-mail 词汇

late delivery 延迟出货
delay the delivery of order 延迟订单出货
unexpected problem 突发问题
delivery process 出货流程
delivery system　出货系统
delivery schedule 出货的时间表，配送时间表

Section Six　Culture Tips

高铁改变全球最大规模的人口流动

每年中国都有几亿人挤上火车，踏上返乡旅途，欢度农历新年。这种拥挤不堪、令人不适的乘车体验，正在因中国大力发展高铁而快速改变。

中国已拥有全球里程最长的高铁路网，尽管如此，中国在未来两年里仍将再投资 3.5 万亿元，使全国铁路路网总长增加 18％，达到 15 万千米。

这项投资大多用于高铁网络的西部拓展建设。沿线地区群山叠嶂，昔日行路难于上青天，难怪古代诗人李白曾因此叹惜。

在春节期间，中国乘坐火车出行的人数将近 4 亿，超过了美国人口总数。中国的工厂和各个单位停工休假一周，从而触发了全球规模最大的人口流动。中国有 14 亿人口，其中大多会返乡与家人团聚，同时也有越来越多的人借此机会在国内外旅游。

自从航空公司推出折扣机票以来，乘坐火车出行的吸引力在世界其他地区已经减弱，但在中国却与日俱增。官方数据显示，2017 年春节创下了单日客流量 1 096 万人次的历史新高，同时高铁客流量首次超过普铁客流量。

十年前中国基本上还没有高铁，但如今中国高铁的发展一日千里，路网规模已达 2.5 万千米，其中半数以上为 2013—2017 年铺设完成的。中国计划到 2025 年将高铁路网规模扩大

50%以上，到2030年年底建成"八横"高铁干线。中国还有意铺设"八纵"干线。

这样，中国高铁网络将会把欠发达的西部地区纳入其中——在建设初期，该网络只是为了在相对富裕的东部沿海地区主要经济中心之间建立起高速连接。

Section Seven Self-evaluation

Rate your learning outcomes in this unit.

Evaluation Grades		Items				
		A	B	C	D	E
Attitude	I can take the initiative to preview before class.					
	I can take an active part in class activities.					
	I can finish tasks carefully and independently.					
Knowledge	I can make good use of words and expressions concerning high speed trains.					
Ability	I can understand the introduction of high speed trains.					
	I can tell the history of high speed trains and the development of Chinese rails.					
	I can write a proper late delivery apology letter.					
Quality	I can understand the strength of China.					
	I am proud of Chinese high speed trains.					
Is there any improvement over the last unit?		YES			NO	

言必信，行必果。　　　　　　　　　　　　　　——《论语》
Promises must be kept; actions must be resolute.　　—*The Analects of Confucius*

Unit Seven

Robots

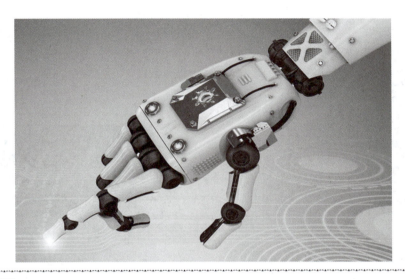

Focus

Section One	Warming Up
Section Two	Dialogue: Do You Like Robots to Work at Your Home?
Section Three	Passage: Industrial Robots
Section Four	Translation Skills
Section Five	Writing: A Business Thank You Letter
Section Six	Culture Tips
Section Seven	Self-evaluation

Learning Objectives

Upon completion of the unit, students will be able to:

1. Have basic knowledge of various kinds of robots.
2. Fully understand the application of industrial robots.
3. Keep in mind the advantages of industrial robots.

Section One Warming Up

Warming Up 听力

Group Work

Match the following types of cleaning robots.

A. Lawn mowing.   B. Pool cleaning.
C. Window cleaning.   D. Robot vacuum cleaner.

Listen to the following material and find out which type of robot is mentioned.

Unit Seven Robots

Section Two — Dialogue: Do You Like Robots to Work at Your Home?

Listen and role-play the following dialogue.

Yang Kang: **Congratulations** on your getting the job, Sally.

Sally: Thank you. But I am quite **exhausted** every day. I am **eager** to enjoy a more **relaxing** life after work.

Yang Kang: Then, you need a **household** service robot. I am **obsessed** with robots. Don't you think it is a magic that they can fully understand the **instructions** we give to them?

Sally: Do you like robots to work at your home?

Yang Kang: Sure. I am pretty **content** with robots working at my home doing things that I really don't like. I can set up the vacuum cleaner to do the **chore**. Multi-function robotic cookers are able to fry, steam, bake, slow cook, and perform any other action.

Sally: Wow, it sounds **awesome**. Loud-connected home robots are already becoming a part of your life. They are super helpers and make life easier.

Yang Kang: It is **essential** that the robots keep your home safe and **secure**. Smart devices can connect with each other. A robot can be the control center for everything from your coffee machine to locking doors. And you can **access** it from anywhere in the world!

Sally: Yes, I can control the automatic pet feeder with my cell phone. A household service robot is the perfect **companion** letting me keep an eye on the cats' condition without having to worry about them.

Words & Expressions

congratulation	[kənˌɡrætʃuˈleɪʃən]	n. 祝贺；恭喜
exhausted	[ɪɡˈzɔːstɪd]	adj. 用完的；精疲力竭的
eager	[ˈiːɡə]	adj. 热切的；渴望的
relaxing	[rɪˈlæksɪŋ]	adj. 轻松的
household	[ˈhaʊshəʊld]	n. 家庭；户
		adj. 家庭的；家喻户晓的
obsessed	[əbˈsest]	adj. 着迷的

instruction	[ɪnˈstrʌkʃn]	n.	指令；教学；说明
content	[ˈkɒntent]	adj.	满足的；愿意的　n. 内容；含量
chore	[tʃɔː]	n.	家务；琐事；讨厌的工作
awesome	[ˈɔːsəm]	adj.	了不起的；精彩的；令人惊叹的
essential	[ɪˈsenʃl]	adj.	本质的；必要的　n. 要素；必需品
secure	[səˈkjʊə]	adj.	安全的；稳妥的　vt. 使安全；固定
access	[ˈækses]	vt.	进入；获取，访问（计算机信息）
		n.	入口；（对计算机存储器的）访问
companion	[kəmˈpæniən]	n.	同伴；志趣相投的人　v. 陪伴

household service robot	家用机器人
vacuum cleaner	吸尘器
automatic pet feeder	自动喂食器
keep an eye on	留意；照看

 Find Information

你喜欢机器人在家里工作吗？

Task Ⅰ. Read the dialogue carefully and then answer the following questions.

1. Does Sally's new job change her life?
_____
2. What advice does Yang Kang give to Sally?
_____
3. Why does Yang Kang like a household robot?
_____
4. How does a robot keep the home safe and secure?
_____
5. Do you think the household service robot is a perfect companion?
_____

Task Ⅱ. Read the dialogue carefully and then decide whether the following statements are true（T）or false（F）.

(　　) 1. Sally can't enjoy a relaxing life after work now.
(　　) 2. Yang Kang often gives instructions to his household robots.
(　　) 3. Yang Kang needn't set up the vacuum cleaner to do the housework.
(　　) 4. Speech-recognition robots are already becoming a part of our life.
(　　) 5. Robots can keep our home safe and secure in some way.

Unit Seven Robots

Cheer up Your Ears

Listen and write down what you've heard. Then read and recite till you can use them fluently.

1. He is eager to enjoy a more _____ life after work.

2. He is _____ with robots because they can fully understand the instructions people give to them.

3. He is _____ with robots doing the housework.

4. You can set up the vacuum cleaner to do the _____ .

5. _____ robotic cookers are able to fry, steam, bake and slow cook.

6. _____ home robots are super helpers and make life easier.

7. It is _____ that the robot can keep your home safe and secure.

8. A robot can be the control center for everything and you can _____ it from anywhere.

9. You can control the _____ pet feeder with a cell phone.

10. A _____ service robot can help people keep an eye on the cats' condition.

Words Building

DIALOGUE CHEER UP EARS

Task Ⅰ. Choose the best answer from the four choices A, B, C and D.

( ) 1. You need a password to get access _____ the computer system.
A. about　　　　B. for　　　　C. to　　　　D. on

( ) 2. They were exhausted _____ the great distances they had walked.
A. by　　　　B. of　　　　C. on　　　　D. because

( ) 3. I was _____ when my child was thrown from the horse.
A. eager　　　　B. anxious　　　　C. worrying　　　　D. careful

( ) 4. A few people were _____ to pay the fines.
A. eager　　　　B. obsessed　　　　C. content　　　　D. satisfied

( ) 5. It is essential that the application forms _____ sent back before the deadline.
A. be　　　　B. are　　　　C. will be　　　　D. were

Task Ⅱ. Fill in each blank with the proper form of the word given.

1. Christie looked _____ and calm as he faced the press. (relax)

2. I send you my warmest _____ on your success. (congratulate)

3. You'll find it _____ to talk things over with a friend. (help)

4. He seemed _____, less bitter. (content)

5. The life of western Hunan _____ him and applied him with many writing materials. (obsess)

Task Ⅲ. Match the following English terms with the equivalent Chinese items.

A. maintenance instructions　　　　1. ( ) 多功能厅

B. access control
C. vacuum cleaner
D. household service robot
E. essential goods
F. awesome achievement
G. mobile device
H. multi-function hall
I. perform magic
J. Intel smart memory access

2. (　　) 家用机器人
3. (　　) 英特尔智能内存访问
4. (　　) 访问控制
5. (　　) 令人惊叹的成就
6. (　　) 必需品
7. (　　) 表演魔术
8. (　　) 移动设备
9. (　　) 吸尘器
10. (　　) 维修说明

 Table Talk

Pair Work

Role-play a conversation about using a household service robot with your partner.

Situation

Imagine you want to buy a household service robot. You weigh up the pros and cons of robots. Now you are talking with your partner.

Section Three　Passage: Industrial Robots

Pre-reading Questions

1. What do you know about industrial robots?
2. How are industrial robots changing the manufacturing industry?
3. Will industrial robots replace human beings in the workplace completely?

An industrial robot is an **automatically** controlled, **reprogrammable**, **multipurpose manipulator**. The most common use of industrial robots is for simple and **repetitive** industrial tasks, including **assembly** line processes, picking and **packing**, **welding**, and similar functions. They offer **reliability**, **accuracy**, and speed. Workers are often exposed to **harsh** or difficult conditions such as heights, **excess** dust, loud noise or **toxic** gases. By introducing robots, efficiency can be improved and the safety of workers is guarded. In the case of **detection** of **defects** in industrial products, traditional manual **inspection** is expensive and inspectors may miss

Unit Seven　Robots

problems or make mistakes, while using machine vision to detect defects can greatly improve efficiency and accuracy. Also, industrial robots have big advantages over people in terms of **standardization**, safety and **intelligence**.

As development progresses, robots are expected to replace even more human work and, in effect, help **replenish** the labor force. Robotics is particularly important to promote the **transformation** from traditional manufacturing to smart manufacturing. Robots have become the core equipment to make smart manufacturing happen and an integral part of smart factories. The robotics industry is an important symbol of a country's technological strength and level of high-end manufacturing.

Words & Expressions

工业机器人

automatically	[ˌɔːtəˈmætɪkli]	adv. 自动地；机械地
reprogrammable	[riːˈprəʊɡræməbl]	adj. 可改编程序的
multipurpose	[ˈmʌltɪˌpɜːpəs]	adj. 多用途的；多目标的
manipulator	[məˈnɪpjuleɪtə]	n. 操纵器；操作者
repetitive	[rɪˈpetətɪv]	adj. 重复的；反复的
assembly	[əˈsembli]	n. 装配；集合
packing	[ˈpækɪŋ]	n. 包装
welding	[ˈweldɪŋ]	n. 焊接
reliability	[rɪˌlaɪəˈbɪləti]	n. 可靠性
accuracy	[ˈækjərəsi]	n. 准确性；精确度
harsh	[hɑːʃ]	adj. 恶劣的；粗糙的；使人不舒服的
excess	[ɪkˈses]	adj. 过量的；额外的 n. 超过；过量
toxic	[ˈtɒksɪk]	adj. 有毒的
detection	[dɪˈtekʃn]	n. 检查；探测
defect	[ˈdiːfekt]	n. 缺点；缺陷
inspection	[ɪnˈspekʃn]	n. 视察；检查
standardization	[ˌstændədaɪˈzeɪʃn]	n. 标准化
intelligence	[ɪnˈtelɪdʒəns]	n. 智能；智力
replenish	[rɪˈplenɪʃ]	vt. 补充；再装满
transformation	[ˌtrænsfəˈmeɪʃn]	n. 转型；转化；改造

loud noise	很大的噪声
in the case of	在……的情况下；就……来说
in terms of	在……方面；就……而言
in effect	实际上；有效
labor force	劳动力
high-end manufacturing	高端制造

 **Find Information**

Task Ⅰ. Read the passage carefully and then answer the following questions.

1. What is the definition of industrial robots?

2. What kind of tasks are industrial robots commonly involved in?

3. What kind of working conditions should be improved for workers?

4. What's the advantage of machine vision when detecting defects in industrial products, compared with manual inspection?

5. What is the key factor that guarantees the transformation from traditional manufacturing to smart manufacturing.

Task Ⅱ. Read the passage carefully and then decide whether the following statements are true (T) or false (F).

(　　) 1. Industrial robots are those numerical control systems that can be automatically controlled, and are reprogrammable, multipurpose manipulators.

(　　) 2. All the industrial robots can be used in assembly line processes, picking and packing, welding, and similar functions.

(　　) 3. It is no doubt that industrial robots are better than factory hands in terms of standardization, safety and intelligence.

(　　) 4. Industrial robots are expected to totally replace human work in manufacturing.

(　　) 5. The robotics industry is a significant mark of a country's technological strength and level of high-end manufacturing.

 **Cheer up Your Ears**

PASSAGE CHEER UP EARS

Listen and write down what you've heard. Then read and recite till you can use them fluently.

1. The most common use of industrial robots is for simple and _____ industrial tasks.

2. Industrial robots can be used in assembly line _____, picking and packing and welding.

3. They offer reliability, _____, and speed.

4. Workers are often _____ to harsh or difficult conditions.

5. _____ can be improved and the safety of workers is guarded by introducing robots.

6. Industrial robots have big advantages over people in terms of _____, safety and intelligence.

Unit Seven　Robots

7. Robots will replace even more human work and, in effect, help replenish the _____.

8. Robotics is particularly important to promote the _____ from traditional manufacturing to smart manufacturing.

9. Robots have become the core equipment to make smart _____ happen and an integral part of smart factories.

10. The robotics industry is an important symbol of a country's _____ strength and level of high-end manufacturing.

Words Building

Task Ⅰ. Translate the following phrases into English or Chinese.

1. 噪声 _____
2. 智能工厂 _____
3. 人工检测 _____
4. 核心设备 _____
5. 补充劳动力 _____
6. high-end manufacturing _____
7. standardization, safety and intelligence _____
8. assembly line processes _____
9. technological strength _____
10. promote the transformation _____

Task Ⅱ. Choose the best answer from the four choices A, B, C and D.

(　　) 1. Was it an accident or did David do it _____?
A. on earth　　　　　　　　B. by choice
C. on purpose　　　　　　　D. by chance

(　　) 2. _____ measurement is very important in science.
A. Repetitive　　　　　　　B. Perfect
C. Excess　　　　　　　　　D. Accurate

(　　) 3. The plane is used for electronic jamming and radar _____.
A. detection　　　　　　　　B. defect
C. direction　　　　　　　　D. distinction

(　　) 4. _____ failure, their position would be dangerous in the extreme.
A. In case of　　　　　　　B. In the case of
C. In terms of　　　　　　　D. If

(　　) 5. We have replaced human labour _____ machines.
A. of　　　　　　　　　　　B. with
C. to　　　　　　　　　　　D. for

127

Task Ⅲ. **Fill in each blank with the proper form of the word given.**

1. Once installed, this heater operates _____. (automatic)
2. They're becoming _____ and getting faster. (programmable)
3. He could also have cut out much of the _____ and thus saved many pages. (repetitive)
4. Friends and neighbors _____ to swap gossip and have drinks. (assembly)
5. Broadcast news was accurate and _____ but deadly dull. (reliability)

Task Ⅳ. **Complete the sentences below with the correct form of the words and phrases in the box.**

| manufacture | labor | intelligence | weld |
| excess | transform | defect | replenish |

1. It's time _____ the shop with the News Year's goods.
2. Maybe the earth has been visited by _____ creatures from outer space.
3. In the two years since I last saw her, her life _____.
4. British _____ industry has been running down for years.
5. Although I have many mistakes and _____, they still like me.
6. The _____ Day in 2030 falls on a Sunday.
7. Make sure that you don't have to pay expensive _____ charges.
8. His job is _____ parts together.

Task Ⅴ. **Translate the following sentences into Chinese.**

1. An industrial robot is an automatically controlled, reprogrammable, multipurpose manipulator.

2. The most common use of industrial robots is for simple and repetitive industrial tasks, including assembly line processes, picking and packing, welding, and similar functions.

3. Workers are often exposed to harsh or difficult conditions such as heights, excess dust, loud noise or toxic gases.

4. As development progresses, robots are expected to replace even more human work and, in effect, help replenish the labor force.

5. The robotics industry is an important symbol of a country's technological strength and level of high-end manufacturing.

Unit Seven   Robots

 **Listening**

Task Ⅰ. Listen to 5 short dialogues and choose the best answer.

1. A. There is no paper.   B. The man can use the printer.
   C. The man has to pay first.   D. The printer doesn't work.
2. A. He has got a pay rise.   B. He has got a new job.
   C. He has been promoted.   D. He has bought an apartment.
3. A. Sell a car.   B. Repair a car.
   C. Rent a car.   D. Buy a used car.
4. A. It is too small.   B. He doesn't like the style.
   C. He doesn't like the color.   D. It is of poor quality.
5. A. The sales manager.   B. The receptionist.
   C. The office secretary.   D. The chief engineer.

Task Ⅱ. Listen to the passage and fill in the blanks with the missing words or phrases.

I think we'll begin now. First I'd like to welcome you all and thank you for your coming, especially at such short notice. I know you are all very busy and it's difficult to take time away from your 1. _____ for meetings. As you can see on the agenda, today we will focus on the upcoming 2. _____. First we'll discuss the groups that will be coming from Germany. After that, we'll discuss the North American tours, 3. _____ by the Asian tours. If time 4. _____, we will also discuss the Australian tours which are booked for early September. Finally, I'm going to request some feedback from all of you 5. _____ last year's tours and where you think we can improve.

 **Extensive Reading**

Directions: *After reading the passage, you are required to choose the best answer to complete the sentences.*

A. job-eating	B. complicated	C. explain	D. intense	E. helpmate
F. strategies	G. instructed	H. executives	I. livelihoods	J. initially

Robot Management

Until now, 1. _____ have largely ignored robots, regarding them as an engineering rather than a management problem. This cannot go on: robots are becoming too powerful and ubiquitous. Companies may need to rethink their 2. _____ as they gain access to these new sorts of workers.

The first issue is how to manage the robots themselves. Asimov laid down the basic rule in 1942: no robot should harm a human. This rule has been reinforced by recent technological improvements: robots are now much more sensitive to their surroundings and can be 3. _____ to avoid hitting people. But the Pentagon's plans make all this a bit more 4. _____: many of its

129

robots will be, in essence, killing machines.

A second question is how to manage the homo side of homo-robo relations. Workers have always worried that new technologies will take away their 5._____. That worry takes on a particularly 6._____ form when the machines come with a human face: Capek's play that gave robots their name depicted a world in which they 7._____ brought lots of benefits but eventually led to mass unemployment and discontent.

So, companies will need to work hard to persuade workers that robots are productivity-enhancers, not just 8._____ aliens. They need to show employees that the robot sitting alongside them can be more of a 9._____ than a threat. Audi has been particularly successful in introducing industrial robots because the car maker asked workers to identify areas where robots could improve performance and then gave those workers jobs overseeing the robots. Employers also need to 10._____ that robots can help preserve manufacturing jobs in the rich world. One reason why Germany has lost fewer such jobs than Britain is that it has five times as many robots for every 10,000 workers.

Section Four Translation Skills

科技英语的翻译方法与技巧——合译法

合译法与拆译法相反，它是将原文中相邻的两个或两个以上的句子紧缩成一个汉语句子，或者将原文中的一个主从复合句译成一个简单句，使译文更加简洁紧凑，意义更加明确和严谨。

A级听力单选

一、简单句的合译

简单句的合译就是把两个或多个相邻的，但形式上又独立的简单句合并译成一个汉语单句或译成一个含有多个分句的句子。

（一）句间有总分关系

相邻的两个句子间存在紧密的总分关系，即前一句是概括性内容，后一句是详细说明；或者前一句是详细的介绍，后一句是评论或结论。针对这种情况，可以将它们合译。翻译实践中，某一句子中往往带有特定单词，如表示范围的among、include、of等介词，或含有this、these等指示代词，或含有example等名词。如：

A级听力填空

With regard to choice and consumption of food, all human sensory perceptions are involved. Among them, vision is the most important one for selecting food and appreciating its quality.

【译文】在选择和品尝食物时，需要使用各种人体感官，其中色泽是选择食物和鉴定质量最为重要的因素。

Any program that might demand precise alignment must offer a fine scrolling facility. This includes, at a minimum, all drawing and painting programs, presentation programs, and image-manipulation programs.

【译文】任何要求精准对其的程序必须提供精准滚动程序，这至少包括各种绘图和绘画程序，文稿演示程序和图像操作程序。

（二）句间有顺接关系

有时相邻句子间存在一种顺接或递进关系，汉译时可合译。这种句子也会有一些特定单词，如表示顺接的 the same is true of (for) 和 and 等，以及表示递进的 too、also、then 等副词。如：

Water pressure is needed to force water along a pipe. Similarly, electrical pressure is needed to force current along a conductor.

【译文】水管中的水流动需要水压，导体中的电流移动同样需要电压。

It starts at the battery when chemical energy is changed into electrical energy. Electrical energy is then changed into heat and eventually into light energy.

【译文】电池中的化学能首先转变成电能，然后变成热能，最后变成光能。

（三）句间有逆接关系

相邻句除了表示顺接关系，还可以表示转折关系。这些句子常含有 but、however、nevertheless、by contrast、on the other hand 等表示转换的连词或副词。汉译时，可以将它们合译，并适当增加一些关联词。如：

Many people think that they're scared of computers because they think they're unfamiliar with them. But that isn't really true.

【译文】许多人因对计算机不熟悉而惧怕使用它们，但事实并非如此。

Errors are generally single, incorrect actions. Exploration, by contrast, is a long series of probes and steps, some of which are keepers and others that must be abandoned.

【译文】错误通常是简单且不正确的行为，而探索与此相反，是一系列的探查步骤。其中一些需要保留，而另外一些需要放弃。

（四）句间有因果关系

英语中表示因果关系的相邻句，有些是前因后果，有些是前果后因，后一句也常有 because、why、so、then、thus、hence、therefore 等词。汉译时，我们可以合译，并适当增加一些汉语的关联词。如：

Pure acetic acid is sometimes called glacial acetic acid. This is because pure acetic acid freezes at 17℃.

【译文】纯醋酸有时也称为冰醋酸，因为它在 17 ℃时就会结"冰"。

The small wire has more resistance. Therefore, less current flows in circuit.

【译文】电线越细，电阻越大，所通过的电流就越少。

（五）句间有条件关系

英语句中除了含 if、given that、as long as 等连词的条件复合句外，相邻句子之间也有用祈使句表示条件，如英文产品说明书，翻译时常用合译法。如：

Drag the program icon to the Start button. The application will then be entered on the main list under that button.

【译文】只要将程序图标拉到"开始"按钮上,就可以将应用程序添加到它的列表上。

To sort your folders and files, click on View, Details. You will see four columns: Name, Size, Type and Date Modified.

【译文】为了对文件夹和文件排序,只要单击"查看"—"详细信息",就会出现四栏信息:名称、大小、类型和修改日期。

二、复合句的合译

英语中的主从复合句包括主语从句、宾语从句、定语从句和状语从句等。这些从句都含有完整的主谓结构。翻译时可根据实际需要将主从复合句中的主句和从句合译。

(一) 含有主语从句的复合句合译

如果 it 引导的主语从句结构简单、短小,可以用合译法。如:

It is not necessary that these products are immediately stored.

【译文】这些产品不需要立即储藏。

It is apparent that bases contain hydroxide ions.

【译文】碱中显然含有氢氧根离子。

(二) 含有宾语从句的复合句合译

有些宾语从句结构简单、句子短,没有必须单独译成一句,可以将它译成一个词组,直接跟在位于后面。如:

Physical chemistry deals with the structure of matter and how energy affects matter.

【译文】物理化学研究物质的结果和能量影响物质的方式。

Choose where you want to save the attachment, then click Save.

【译文】首先选择保存文件的位置,然后单击"保存"。

(三) 含有定有从句的复合句合译

英语限制性定语从句除用"换序法"翻译外,也可以用"合译法"翻译。如:

Double-click the My Computer icon which is displayed on the desktop.

【译文】双击桌面上"我的电脑"图标。

Now choose the files and folders options that suit your needs.

【译文】选择你需要的文件和文件夹选项。

(四) 含有状语从句的复合句合译

英语中结果简单、短小的状语从句,可以翻译成一个短语,然后再与其他成分合并组成一个简单句。如:

When you access the Internet, the connection (log-on) program performs the following tasks.

【译文】联网的应用程序能够执行如下任务。

The liquid water is heated so that it becomes steam.

【译文】液态水受热会变成蒸汽。

Unit Seven　Robots

Section Five　Writing：A Business Thank You Letter

感谢下订单/合作

一、写作要点 Key Points

Step 1：感谢对方的订单/合作。
Step 2：承诺完成订单，或确认订单内容。
Step 3：表示希望未来能继续合作。

二、实际 E-mail 范例

| 写信▼ | 删除 | 回复▼ | 寄件者： | Ginny |

Dear Sir/Madam,

Introduction
We are pleased that you have chosen JBS for your bathroom equipment purchase. JBS is delighted to welcome you as our new client.

Body
We hope you are satisfied with the excellent quality of your new bathtub.
Please be kindly advised that all our products have two years' guarantee from the date of purchase.

Closing
Should you have any questions concerning our products, please call 1234-5678 for our customer service, which is available 24/7.
Thank you again for your purchase.
We look forward to serving you again.

Best regards,
Signature of Sender/Sender's Name Printed

自动控制英语

| E-mail 中译 | 写信▼ | 删除 | 回复▼ | 寄件者： | Ginny |

您好：

开头
很高兴您决定选购 JBS 的卫浴设备产品，我们很开心欢迎您成为我们的新客户。

本体
我们希望您很满意新浴缸的良好质量。
在此贴心提醒您，我们所有的产品都包含自购买日起 24 个月的保修期。

结尾
万一您对本公司产品有任何问题，请拨打本公司 24 小时全年无休的客服专线 1234-5678。再次感谢您的购买。
期待能再次为您服务。

敬祝安康，
签名档/署名

三、各段落超实用句型

说明：画下划线部分的单词可按照个人情况自行替换。

（一）开头 Introduction

❶Thank you very much for ordering clothes and accessories from us.
非常感谢您向我们订购服饰与配件。
❷We are pleased that you have chosen our products for your office equipment purchase.
很高兴您选择本公司产品作为您的办公室设备。
❸We are very delighted to welcome you as our new client.
我们很高兴欢迎您成为本公司的新客户。
❹We have received your order. Thanks you for choosing our products.
我们已经收到您的订单了。感谢您选择我们的产品。
❺Thanks for ordering our products from our websites.
感谢您在本公司网站上订购产品。

（二）本体 Body

❶We hope you are happy with the excellent quality of your new washing machine.
我们希望您对您选购的新洗衣机的良好质量感到很满意。
❷We hope you have as much fun using it as we did designing it.
我们希望您在使用这项产品时，就如同我们在设计它时一样开心。
❸Please be kindly advised that all our products have two years' guarantee from the date of

Unit Seven Robots

purchase.

在此贴心通知您，我们所有的产品都有自购买日起<u>两年</u>的保修期。

❹Please note that your order will be sent to you <u>COD</u> within <u>five working days</u>.

在此通知您，您订购的商品会在<u>五个工作日</u>内以<u>货到付款的方式</u>寄送给您。

❺We are pleased to inform you that your order has been processed and will be delivered to you upon receipt of your payment.

很高兴通知您，我们已经受理您的订单，并且在收到您的订单款项后就会立刻寄送给您。

（三）结尾 Closing

❶We appreciate the opportunity to do business with you.

我们感谢有与您合作的机会。

❷We look forward to a long and mutually beneficial relationship with you.

我们期待与贵公司建立长久的互惠合作关系。

❸If you are interested in any other commodities, please let me know.

如果您对本公司其他产品有兴趣，请与我联络。

❹Should you have any questions concerning our products, please contact our 24/7 customer service.

万一您对本公司产品有任何问题，请拨打本公司24小时全年无休的客服部联络专线。

❺We are grateful for your patronage and hope you will continue to do business with us in the future.

我们很感谢您的惠顾，希望您未来也会持续与我们合作。

Section Six Culture Tips

机器人的发展历程

机器人按其发展历程可以分为以下三代。

第一代的机器人是遥控操作的机器，不能离开人的控制而独自运动，是通过一台计算机控制一个多自由度的机器，只能执行简单、重复性的任务，如生产线上的装配工作。这些机器人的智能程度较低，需要依靠外部的指令或程序来执行任务，对周围环境没有感知与反馈控制的能力。第一代机器人也称为示教再现型机器人。

第二代机器人是基于计算机和传感器的智能机器人，可以感知周围的环境和状态，并做出相应的决策和行动。这代机器人在工作时，根据感觉器官（传感器）获得的信息，灵活调整自己的工作状态，以保证在适应环境的情况下完成工作。例如，有触觉的机械手可轻松自如地抓取鸡蛋，具有嗅觉的机器人能分辨出不同的饮料和酒类。这些机器人的智能程度较高，但还需要依靠外部的指令或程序来执行一些复杂的任务。

第三代机器人是具有高度自主的智能机器人，不仅具有比第二代机器人更完善的环境感知能力，而且还具有逻辑思维、判断和决策的能力，可根据作业要求与环境信息自主地进行工作。第三代机器人先利用各种传感器、测量器等来获取环境信息，然后利用智能技术进行识别、理解、推理，最后作出决策和规划，是能自主运动实现预定目标的高级机器人。这代机器人已经具有了自主性，有自行学习、推理、决策、规划等能力，其未来发展方向是有知觉、有思维、能与人对话。

Section Seven　Self-evaluation

Rate your learning outcomes in this unit.

Evaluation Grades		Items				
		A	B	C	D	E
Attitude	I can take the initiative to preview before class.					
	I can take an active part in class activities.					
	I can finish tasks carefully and independently.					
Knowledge	I can make good use of words and expressions concerning robots.					
Ability	I can read and understand the reading material.					
	I can describe the application of robots.					
	I can write a business thank you letter.					
Quality	I can understand the importance of robots.					
	I can talk about robots with others.					
Is there any improvement over the last unit?		YES			NO	

修身、齐家、治国、平天下。　　　　　　　　——《礼记》
Self-cultivation, family regulation, state governance, bringing peace to all under heaven.
—*The Book of Rites*

Unit Eight

A Smart City

Focus

Section One	Warming Up
Section Two	Dialogue: What Does a Smart City Look Like?
Section Three	Passage: Smart Cities and Normal Cities—What Are the Differences?
Section Four	Translation Skills
Section Five	Writing: A Job Promotion Congratulation Letter
Section Six	Culture Tips
Section Seven	Self-evaluation

Learning Objectives

Upon completion of the unit, students will be able to:

1. Have basic knowledge of a smart city.
2. Fully understand the differences between a smart city and a normal city.
3. Keep in mind the advantages of a smart city.

Section One   Warming Up

Group Work

Do you know the following famous smart cities in the world?

A. Helsinki Finland.   B. Singapore.   C. Barcelona Spain.

_____

_____

Unit Eight　A Smart City

Warming Up 听力

Listen to the dialogue and find out which smart city above they are talking about.

Section Two　Dialogue: What Does a Smart City Look Like?

Listen and role-play the following dialogue.

(Yang Kang and Sally are at the China International Smart City Expo World Congress)

Sally: Would you like to live in a city where buildings turn the lights off for you, where self-driving cars find the nearest parking space, and where even the rubbish bins know when they're full?

Yang Kang: Of course. But what exactly does a smart city look like?

Sally: A smart city refers to a **metropolitan** or **cosmopolitan** city that **utilizes** internet of things technology to effectively manage the city's assets and resources.

Yang Kang: What are the **characteristics** that define a smart city?

Sally: Each smart city will have their own unique objective, but they all acknowledge the importance of IOT technology and understand that emerging IOT and cloud capabilities offer a meaningful **opportunity** to better understand the **intimate** workings of an **urban** center.

Yang Kang: How can cities become smart?

Sally: The key for a city to become smart lies in how it can **leverage** both IOT technology and location-awareness.

Yang Kang: Can any city become smart?

139

Sally: Any city can become a smart city, regardless of size, as long as it lays the key supporting **infrastructure** and has the **commitment** of citizens, **municipal** officials and local government and **collaboration** between public and private companies with the different levels of government.

Words & Expressions

metropolitan	[ˌmetrəˈpɒlɪtən]	adj. 大都市的
cosmopolitan	[ˌkɒzməˈpɒlɪtən]	adj. 世界性的
utilize	[ˈjuːtəlaɪz]	vt. 利用或使用
characteristic	[ˌkærəktəˈrɪstɪk]	n. 特点；特性；特色
opportunity	[ˌɒpəˈtjuːnəti]	n. 机会，时机
intimate	[ˈɪntɪmət]	adj. 亲密的，密切的
urban	[ˈɜːbən]	adj. 城市的
leverage	[ˈliːvərɪdʒ]	vt. 对……产生影响
infrastructure	[ˈɪnfrəstrʌktʃə(r)]	n. 基础建设，公共建设
commitment	[kəˈmɪtmənt]	n. 信奉，献身
municipal	[mjuːˈnɪsɪpl]	adj. 市的，市政的
collaboration	[kəˌlæbəˈreɪʃn]	n. 合作，协作

refer to...	指的是……
regardless of	不管，不顾，不理会
as long as	只要
lie in	取决于

GO Find Information

Task Ⅰ. Read the dialogue carefully and then answer the following questions.

1. What exactly does a smart city look like?

2. What are the characteristics that define a smart city?

3. How can cities become smart?

4. Can any city become smart?

Unit Eight　A Smart City

5. Is there a city where buildings turn the lights off for you, where self-driving cars find the nearest parking space, and where even the rubbish bins know when they're full?

Task Ⅱ. Read the dialogue carefully and then decide whether the following statements are true (T) or false (F).

(　　) 1. There isn't a city where buildings turn the lights off for you, where self-driving cars find the nearest parking space, and where even the rubbish bins know when they're full.

(　　) 2. A smart city refers to a metropolitan or cosmopolitan city that utilizes internet of things (IOT) technology to effectively manage the city's assets and resources.

(　　) 3. IOT is the short for internet of things.

(　　) 4. The key for a city to become smart lies in the size of it.

(　　) 5. If the city can leverage both IOT technology and location-awareness, it can be called a smart city.

Cheer up Your Ears

DIALOGUE CHEER UP EARS

Listen and write down what you've heard. Then read and recite till you can use them fluently.

1. Would you like to live in a city where _____ turn the lights off for you?

2. There self-driving cars find the nearest _____ .

3. This is the city where even the _____ know when they're full.

4. But what exactly does a smart city _____ ?

5. A smart city _____ a metropolitan or cosmopolitan city that utilizes internet of things technology to effectively manage the city's assets and resources.

6. What are the characteristics that _____ a smart city?

7. Each smart city will have their own _____ objective.

8. The key for a city to become smart _____ how it can leverage both IOT technology and location-awareness.

9. Any city can become a smart city, regardless of _____ .

10. Emerging IOT and cloud capabilities _____ a meaningful opportunity to better understand the intimate workings of an urban center.

Words Building

Task Ⅰ. Choose the best answer from the four choices A, B, C and D.

(　　) 1. Would you like _____ special for your mother's birthday?

A. to buy something　　B. buying something　　C. to buy anything　　D. buying anything

(   ) 2. What I have to say _____ all of you.

    A. to refer to   B. refer to   C. referring to   D. refers to

(   ) 3. _____ danger, he climbed the tower.

    A. Regard of   B. Regardless   C. Regardless of   D. Regardless to

(   ) 4. _____ I live, I will help you.

    A. As soon as   B. As long as   C. As far as   D. As quickly as

(   ) 5. The cure for stress lies in _____ to relax.

    A. to learn   B. learning   C. learn   D. learnt

Task Ⅱ. Fill in each blank with the proper form of the word given.

1. In order _____ land more fully, they adopted close planting. (utilize)

2. I'll take the job _____ of the pay. (regard)

3. We actively press ahead with the _____ development plan. (structure)

4. He wrote the book in _____ with his colleagues. (collaborate)

5. Ambition is _____ of all successful businessmen. (character)

Task Ⅲ. Match the following English terms with the equivalent Chinese items.

A. waste management   1. (   ) 数字信息亭

B. public safety   2. (   ) 公共安全

C. digital information kiosk   3. (   ) 空气质量传感器

D. intelligent street lamp   4. (   ) 智能路灯

E. parking sensors   5. (   ) 气候监测

F. air quality sensors   6. (   ) 停车传感器

G. climate monitoring   7. (   ) 废物管理

 **Table Talk**

Pair Work

Role-play a conversation about differences between smart cities and normal ones with your partner.

Situation

Imagine Yang Kang and Sally are at the Smart City Expo World Congress. They are talking about the differences between smart cities and normal ones.

Unit Eight   A Smart City

Section Three   Passage: Smart Cities and Normal Cities—What Are the Differences?

Pre-reading Questions

1. What do you know about smart cities?
2. What makes smart cities different from normal cities?
3. What is the main purpose of smart cities?

智慧城市与普通城市有何区别?

Smart cities are a hot **buzzword** right now as cities all over the world look to how they can use new technologies. These cities are no longer **futuristic scenarios** dreamed up by creative thinkers. Instead, real places around the globe are discovering **innovative** ways to incorporate smart technology into people's everyday lives. Let's take a close look at a smart city and a normal city. What are the differences between them? In other words, what makes a smart city smart?

The **conceptualization** of smart city may vary from city to city and country to country depending on the level of development. But simply put, a smart city is an urban area that uses information and communication technologies (ICT) to ease up the livelihood of its people. It is a **municipality** that uses ICT to **augment operational** efficiency, share information with the public and improve both the quality of government services and citizen **welfare**.

The main purpose of smart city is to create a society which can **perform** effectively and efficiently making effective use of city **infrastructures** through **artificial** intelligence. It also focuses to **optimize** city functions and drive economic growth while improving quality of life for its citizens using smart technology and data analysis.

Words & Expressions

buzzword	[ˈbʌzwɜːd]	n. 流行词
futuristic	[ˌfjuːtʃəˈrɪstɪk]	adj. 未来的
scenario	[sɪˈnɑːriəʊ]	n. 情节；情况
innovative	[ˈɪnəveɪtɪv]	adj. 创新的；革新的
conceptualization	[kənˌseptjuəlaɪˈzeɪʃn]	n. 概念化
ease	[iːz]	vi. 减轻；放松

municipality	[mjuːˌnɪsɪˈpæləti]	n. 自治市；市当局
augment	[ɔːgˈment]	vt. 增加
operational	[ˌɒpəˈreɪʃənl]	adj. 操作的；运作的
welfare	[ˈwelfeə(r)]	n. 福利；福利事业
perform	[pəˈfɔːm]	vt. 执行；履行
infrastructure	[ˈɪnfrəstrʌktʃə(r)]	n. 基础设施；基础结构
artificial	[ˌɑːtɪˈfɪʃl]	adj. 人造的
optimize	[ˈɒptɪmaɪz]	vt. 使完善；使优化

dream up	凭空想出
in other words	换句话说
vary from	不同于
depend on	依赖于
simply put	简言之，简单讲，简单说
ease up	减轻，缓和
share with...	与……分享
make use of	利用

GO Find Information

Task Ⅰ. Read the passage carefully and then answer the following questions.

1. What is a hot buzzword right now as cities all over the world look to how they can use new technologies according to the passage?

_____

2. Are there real places around the globe that are discovering innovative ways to incorporate smart technology into people's everyday lives?

_____

3. What is the main purpose of smart city?

_____

4. What can be called a smart city?

_____

5. What does a smart city focus to do?

_____

Task Ⅱ. Read the passage carefully and then decide whether the following statements are true (T) or false (F).

(　　) 1. ICT is the short form of information and communication technologies.

(　　) 2. Smart cities are still futuristic scenarios dreamed up by creative thinkers.

(   ) 3. There is no difference between a smart city and a normal city.

(   ) 4. The main purpose of smart city is to create a society which can perform effectively and efficiently making effective use of city infrastructures through artificial intelligence.

(   ) 5. A smart city also focuses to optimize city functions and drive economic growth while improving quality of life for its citizens using smart technology and data analysis.

Cheer up Your Ears

Listen and write down what you've heard. Then read and recite till you can use them fluently.

1. Smart cities are a hot _____ right now all over the world.

2. These cities are _____ futuristic scenarios dreamed up by creative thinkers.

3. Instead, real places around the globe are discovering innovative ways to incorporate smart technology into people's _____ .

4. The conceptualization of smart city may _____ city to city and country to country depending on the level of development.

5. A smart city is an urban area that uses information and communication technologies to _____ the livelihood of its people.

6. A smart city share information with the public and improve both the quality of government services and citizen _____ .

7. The main purpose of smart city is to create a _____ which can perform effectively and efficiently making effective use of city infrastructures through artificial intelligence.

8. It also focuses to optimize city functions and drive _____ while improving quality of life for its citizens using smart technology and data analysis.

9. Let's take a _____ at a smart city and a normal city.

10. It is a municipality that uses ICT to _____ operational efficiency.

Words Building

PASSAGE CHEER
UP EARS

Task Ⅰ. Translate the following phrases into English or Chinese.

1. 人工智能　　　　　　　　　_____
2. 未来场景　　　　　　　　　_____
3. 日常生活　　　　　　　　　_____
4. 发展水平　　　　　　　　　_____
5. 市民福利　　　　　　　　　_____
6. optimize city functions　　_____
7. improve the quality　　　 _____

8. operational efficiency　　　　_____

9. ease up the livelihood of people　　_____

10. city infrastructures　　　　_____

Task Ⅱ. Choose the best answer from the four choices A, B, C and D.

(　　) 1. The students vary _____ one another in character.

A. from　　　　B. to　　　　C. in　　　　D. on

(　　) 2. This is the kind of atmosphere we want to _____.

A. creative　　　　B. create　　　　C. creation　　　　D. creator

(　　) 3. He is in his _____ clothes.

A. every day　　　　B. every-day　　　　C. everyday's　　　　D. everyday

(　　) 4. Their labor _____ is very high.

A. efficiency　　　　B. efficient　　　　C. efficiencies　　　　D. efficiently

(　　) 5. Smart cities are _____ futuristic scenarios dreamed up by creative thinkers.

A. no longer　　　　B. no long　　　　C. no longest　　　　D. not any longest

Task Ⅲ. Fill in each blank with the proper form of the word given.

1. You are _____ put the cart before the horse. (simple)

2. He gave a lecture on the _____ of solar energy. (useful)

3. She welcomed us with an _____ smile on her face. (artifice)

4. We should emphasize that the _____ process must take into account the fact. (optimize)

5. These cities are no longer futuristic scenarios dreamed up by _____ thinkers. (create)

Task Ⅳ. Complete the sentences below with the correct form of the words and phrases in the box.

dream up	in other words	vary from	depend on
simply put	ease up	share with	make good use of

1. This is the most fantastic rumor one could _____.

2. You should _____ on the child and stop scolding her.

3. Everything _____ the sun for their growth.

4. They asked him to leave—_____ he was fired.

5. —What information will people not _____ each other?

　　—Age and salary.

6. The fish _____ one another in size.

7. _____, software is at the root of all common computer security problems.

8. We will _____ the opportunity offered by the company.

Task Ⅴ. Translate the following sentences into Chinese.

1. The conceptualization of smart city may vary from city to city and country to country depending on the level of development.

2. But simply put, a smart city is an urban area that uses information and communication technologies to ease up the livelihood of its people.

3. It is a municipality that uses ICT to augment operational efficiency, share information with the public and improve both the quality of government services and citizen welfare.

4. The main purpose of smart city is to create a society which can perform effectively and efficiently making effective use of city infrastructures through artificial intelligence.

5. It also focuses to optimize city functions and drive economic growth while improving quality of life for its citizens using smart technology and data analysis.

 **Listening**

Task Ⅰ. Listen to 5 short dialogues and choose the best answer.

1. A. The brand image.　　　　　　B. The marketing strategy.
　C. The sales plan.　　　　　　　D. The company culture.
2. A. Telephone bills.　　　　　　B. Online shopping.
　C. Telephone banking.　　　　　D. Credit cards.
3. A. On the third floor.　　　　　B. On the fifth floor.
　C. On the sixth floor.　　　　　D. On the eighth floor.
4. A. She doesn't like the new house.　B. She can't help the man.
　C. She will go to the concert.　　D. She will be away on business.
5. A. Write a report.　　　　　　　B. Book a flight.
　C. Attend a meeting.　　　　　　D. Meet an engineer.

A级听力单选

A级听力填空

Task Ⅱ. Listen to the passage and fill in the blanks with the missing words or phrases.

Good afternoon passengers. This is the pre-boarding announcement for flight 89B to Moscow. We are now 1. _____ those passengers with small children, and any passengers requiring special assistance, to begin boarding at this time. Please have your 2. _____ and identification ready. Regular boarding will begin in approximately ten minutes' time. Thank you.

This is the final boarding call for passengers Eric and Fred Collins booked on flight 89B to Moscow. Please proceed to 3. _____ immediately. The final checks are being 4. _____ and the captain will order for the doors of the aircraft to close in approximately five minutes' time. I 5. _____. This is the final boarding call for Eric and Fred Collins. Thank you.

 Extensive Reading

Directions: After reading the passage, you are required to choose the best answer from the four choices for each statement.

Smart Cities Aren't Just a Dream of the Future Anymore

With intelligent systems and new-age transit networks, life in the big cities will likely be happier and more efficient.

After all, more than 60% of the world's population is expected to live in cities by 2050, according to a U.N. report. The answer to making these cities more livable for so many people lies in creating smart cities.

These cities will use 5G networks and the internet of things to make everyday life safer and more convenient. Cities like Boston, Baltimore, Amsterdam and Copenhagen are already using smart technology to improve the lives of residents.

But what exactly does a smart city do? Let's look at a few examples.

In the U.S. cities of Boston and Baltimore, smart trash cans can sense how full they are and inform cleaning workers when they need to be emptied. In Amsterdam, the Netherlands, traffic flow and energy usage are monitored and adjusted according to real-time data gathered from sensors around the city. And in Copenhagen, Denmark, a smart bike system allows riders to check on air quality and traffic congestion as they ride.

Smart cities will be interactive, allowing their residents to feel like they're truly shaping their environment, instead of merely existing in it. "One of the most important reasons to have a smart city is that we can actually communicate with our environment in a way that we never have in the past", Mrinalini Ingram, head of the "smart city" initiative at telecom company Verizon, told Tech Republic.

Smart cities will also allow us to save resources. By using sensors and 5G networks to monitor the use of water, gas and electricity, city managers can figure out how to distribute and save these resources more efficiently. Emissions of carbon dioxide and other air pollutants can be more closely monitored in smart cities as well.

Of course, it will take time and money to turn our current cities into the smart cities of the future. But as we've already seen, more cities around the world are already adopting smart technology in small ways. China, for instance, is making investments in big cities like Shanghai and Guangzhou to make them "smarter". It won't be long until even more cities start to develop their own smart infrastructure.

1. What do we know about smart cities?

A. There are more people living in smart cities than in villages.

B. People all over the world will live in smart cities in 2050.

Unit Eight　A Smart City

C. Technology plays a great role in making smart cities a reality.

D. Smart cities, like Boston in the U. S., make full use of 5G networks.

2. What do paragraphs 6 – 7 mainly talk about?

A. How to build smart cities.　　　　B. How smart cities harm animals.

C. How smart cities make life easier.　D. How smart cities help the environment.

3. What is the author's attitude toward smart cities?

A. Doubtful.　　　B. Critical.　　　C. Positive.　　　D. Indifferent.

Section Four　Translation Skills

科技英语翻译技巧——还原法

英语中为了避免重复，往往采用省略或者替换。汉译时，为了尽量使信息完整、清晰，需将英语中省略或替换的成分还原。

一、名词的还原

（一）省略的名词还原

英语句子中为避免重复而省略的名词，在汉译时，为避免译文句式不规范、语义不清楚的情况，必须把省略或替换的名词还原出来。如：

Though both roller and ball bearings serve the same basic purpose, their differences lie in their design and load bearing capacities.

【译文】尽管滚柱轴承和滚珠轴承功能相同，但是它们在设计和承载力方面存在差异。

Scheele discovered a number of organic acids including tartaric, oxalic, uric, lactic, and citric.

【译文】舍勒发现了诸多有机酸，包括酒石酸、草酸、尿酸、乳酸和柠檬酸。

Recall that there are two basic types of electronic signals—analog and digital.

【译文】电子信号有两种：模拟信号和数字信号。

Energy exists in many forms, such as electrical, mechanical, chemical, and heat.

【译文】能量的存在形式有多种，如电能、机械能、化学能和热能。

（二）替换的名词还原

英语中被替换的名词，在翻译时需要还原所替换的部分。如：

While small generators frequently have revolving armatures, large machines usually have stationary armatures and revolving field.

【译文】通常，小型发电机采用旋转的电枢，大型发电机则采用固定的电枢和旋转的激磁绕组。

For many applications, thin-film resistors and capacitors often have certain distinct advantages over their junction counterparts.

【译文】在许多场合，薄膜电阻、薄膜电容与结电阻、结电容相比有一些明显的优点。

FET amplifiers, like their bipolar counterparts, can be connected in three different circuit configurations.

【译文】与双极放大器一样，场效应放大器也可以连接到三种不同配置的线路中去。

在上面的第一个句子中 machines 替换的是 generators，若直译为机器，则易引起误解，所以把其还原翻译成发电机。同样，在第二句和第三句中，若直译 counterparts，也非常不妥。应厘清它所替换的名词，然后翻译。

二、代词的还原

英语中为避免名词重复，除了通过省略或其他名词替换外，还可以通过代词替换。可以替换的代词包括普通代词如 it、they、their、them 等人称代词和物主代词；不定代词如 one、the one、ones 等；指示代词如 that、those、this、these 等；关系代词如 which、that 等。而汉语往往重复名词。因此，英语中这些所替换的代词在汉译时需还原。

（一）普通代词的还原

A diode can be used as a rectifier because in it the current flows in one direction.

【译文】二极管可用作整流器，因为电流在二极管中只朝一个方向流动。

Before a gas discharge tube is sealed, most of the air is pumped out of it.

【译文】在密封排气管之前，先要将管中的大部分空气排出。

All the metals are good conductors because there is a great number of free electrons in them.

【译文】一切金属都是良导体，因为金属中有大量的自由电子。

（二）关系代词的还原

Needs are the basic, often instinctive, human forces that motivate a person to do something.

【译文】各种需要是人类基本的，通常又是出于本能的驱动力，这种驱动力促使一个人去做事。

Vectors are straight lines that have a specific direction and length.

【译文】向量是一些直线，它们有特定的方向和长度。

Another challenging analytical technique is attenuated total reflectance (ATR) spectroscopy, which has been successfully applied in the IR spectral region and also in the visible region.

【译文】另一种富有挑战的分析技术是衰减全反射光谱法，它已经被成功地应用于红外线和可见光光谱领域。

三、动词的还原

（一）省略的动词还原

在以 and、but、or 等词连接或者没有连词的并列句中，后半部分省略与第一分句相同的成分。翻译时常把省略的部分还原。如：

The physical state has changed, but not the substance.
【译文】物质的物理状态发生了变化，但是物质本身并没有变化。
Power considers not only the work that is performed but the amount of time in which the work is done.
【译文】功率不仅考虑做功的多少，还考虑做功所需的时间。
TV cable offers not only better, more reliable signals, but also hundreds of channels of programming.
【译文】电视电缆不仅能提供更好和更可靠的信号，还能提供数百个不同的电视频道。

（二）替换的动词还原

为了避免重复使用同一个动词，英语句子中常用动词 do 或不定式 to 替代前面出现过的动词。汉语中没有类似的用法，因此翻译时要还原替换的动词。

Double-clicking the title bar maximizes a window just as pressing the maximize button does.
【译文】就像单击"放大"按钮能放大窗口一样，双击标题栏也能实现窗口的最大化。
Carotenoids absorb wavelengths of blue light which chlorophylls do not.
【译文】类胡萝卜素能吸收蓝光波长的光能，而叶绿素则无法吸收这种波长的光。
The SMx and PCM8/ST models include non-electrostatic effects; the other models do not.
【译文】SMx 和 PCM8/ST 模型包含了非静电效应，而其他模型没有包含这些效应。

Section Five　Writing: A Job Promotion Congratulation Letter

祝贺职位升迁

一、写作要点 Key Points

Step 1：恭贺对方获得升迁。
Step 2：赞扬对方工作表现终于实至名归。
Step 3：期许对方未来工作表现，并祝福对方未来事业成功。

二、实际 E-mail 范例

| 写信▼ | 删除 | 回复▼ | 寄件者： | Ginny |

Dear Sir/Madam,

Introduction
I am so happy to hear about your promotion to the Chief Engineer position. Please accept my congratulations on your promotion!

Body
I have always been impressed by your outstanding performance at work and your brilliant leadership. I must say that I was not surprised at your promotion at all because this recognition of your ability is well deserved.

Closing
I am already looking forward to hearing about your promotion celebration party. Congratulations again, and wish you continued success in your new position.

Best regards,
Signature of Sender/Sender's Name Printed

E-mail 中译 | 写信▼ | 删除 | 回复▼ | 寄件者： | Ginny

您好：

开头
　　得知您被晋升为总工程师，我非常开心！请接受我对您晋升的祝贺之意！

本体
　　我一直都很欣赏您在工作上的杰出表现以及您优异的领导才能。我必须说我对您的升迁一点都不感到意外，因为您的工作能力值得肯定。

结尾
　　我已经开始期待您的晋升庆祝派对了。
　　再次恭喜您，并祝您在新职位上大展宏图。

敬祝安康，
签名档/署名

Unit Eight　A Smart City

三、各段落超实用句型

说明：画下划线部分的单词可按照个人情况自行替换。

（一）开头 Introduction

❶We are glad to know that you have been promoted to <u>the Manager of the Research and Development Department</u>.
得知您已经被晋升为<u>研发部经理</u>，我们感到非常高兴。

❷It's a thrill to hear that you have been appointed the <u>Manager of the Marketing Department</u>.
听到您已经被任命为<u>营销部经理</u>，真是令人感到兴奋。

❸Please accept my heartiest congratulations on your promotion as <u>the District Manager</u>.
请接受我诚心的祝贺，恭贺您荣升<u>区经理</u>。

❹I am writing this letter to congratulate on your promotion.
我写这封信是要恭贺你升职。

❺Big congratulations on your recent promotion!
恭喜您近日荣获升迁！

（二）本体 Body

❶Getting a promotion is definitely a big career milestone.
获得晋升绝对是事业上一个重要的里程碑。

❷Your dedication at work definitely deserves this recognition.
你对工作的付出绝对值得这样的肯定。

❸I have always been impressed by the high quality of your work, so I wasn't surprised at your promotion at all.
我一直很欣赏你高质量的工作，所以我对你的晋升一点都不感到意外。

❹I am happy for your achievement and you truly deserve the promotion.
我为您的成就感到高兴，此次晋升，您确实当之无愧。

❺Your outstanding performance at work is impressive. This recognition of your ability is well deserved.
你杰出的工作表现让人欣赏，你的工作能力值得这份肯定。

❻I'm delighted to hear that you have received the promotion you deserve and I'm positive that you will play your new role perfectly.
我很高兴听说您获得升迁，这是您应得的，我相信您在新的职位上肯定会做得很好。

（三）结尾 Closing

❶The promotion deserves a fantastic person like you.
这个晋升机会理当给予一个像您这样杰出的人。

❷It was the promotion that you rightfully deserved.

153

这次晋升对你而言是理所当然的。

❸I look forward to being invited to your promotion celebration party.

我期待能受邀参加您的晋升庆祝派对。

❹Wish you continued success in your new position.

祝您在新职位上大展宏图。

❺I am confident that your department will continue growing under your leadership.

我相信贵部门在您的带领下会继续茁壮成长。

❻I wish you continued success in your future endeavors and a lucrative career ahead of you.

祝您未来的事业成功，职业生涯大有收获。

四、不可不知的实用 E-mail 词汇

promotion 晋升

promotion celebration party 晋升庆祝派对

career milestone 事业上的里程碑

recognition 肯定

outstanding performance at work 工作上的杰出表现

high quality of work 高质量的工作

brilliant leadership 优异的领导能力

hard work 认真工作

dedication at work 对工作的投入

well-deserved 应得的，理所当然的

Section Six　Culture Tips

新冠疫情凸显智慧城市优势

　　智慧城市不再只是一个概念而已。实际上，智慧城市应用程序已表明它们可以帮助中国遏制新型冠状病毒传播，促使经济从2020年第一季度的收缩转为第二季度的增长，这是疫情的一个意外收获。在新冠疫情期间，我们在某些城市看到了一些智慧城市的应用程序；随后，它们也在中国其他地区出现了。

　　在中国，防疫智慧系统被称为健康码，有红色、黄色或绿色标识，用于识别一个人是否具有感染新型冠状病毒的高风险性。

　　健康码最初是一款极为流行的支付应用程序上的一项功能，仅为该程序的中国员工使用，后来发展成为某些流行应用程序的国家标准功能。该应用程序的新款防疫功能可以计算一个人是否与新型冠状病毒感染病例接触或聚集过，由此显示其是否可以进入公共空间，比如进入生鲜市场或乘坐公共交通工具。计算的输入源来自街道和智能路灯上安装的传感器和

监控摄像头。健康码不仅可以识别疑似感染病例，还避免了可能造成巨大经济损失的过度社交距离措施；在使用健康码的城市，感染率因此而降低。

  我们"粗略估算"，到 2030 年，北京、上海、广东、深圳和杭州将成为经验丰富的智慧城市技术的实践者。到 2035 年，中国大部分地区都会具备一些智慧城市功用。这意味着中国经济可以更快地实现城市化。

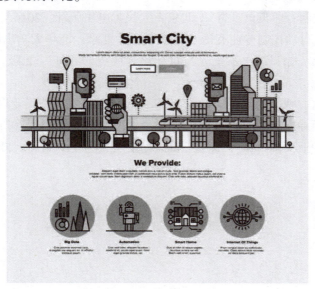

Section Seven Self-evaluation

Rate your learning outcomes in this unit.

Evaluation Grades		Items				
		A	B	C	D	E
Attitude	I can take the initiative to preview before class.					
	I can take an active part in class activities.					
	I can finish tasks carefully and independently.					
Knowledge	I can make good use of words and expressions concerning smart cities.					
Ability	I can understand what a smart city is.					
	I can tell the difference between a smart city and a normal one.					
	I can write a proper job promotion congratulation letter.					
Quality	I can understand the importance of a city's development.					
	I can make contribution to our city construction.					
Is there any improvement over the last unit?		YES		NO		

参 考 文 献

［1］ 李东林. 自动控制专业英语［M］. 哈尔滨：哈尔滨工业大学出版社，2021.
［2］ 自动控制英语教材编写组. 自动控制英语［M］. 北京：高等教育出版社，2017.
［3］ 王晓琳. 自动控制专业英语［M］. 北京：人民邮电出版社，2016.
［4］ 李国厚，黄河. 自动控制专业英语［M］. 2版. 北京：清华大学出版社，2013.
［5］ 龚育尔，代小艳，徐岚. 电器工程与自动化专业英语［M］. 2版. 北京：机械工业出版社，2012.
［6］ 王明文，王璐欢. 智能控制技术专业英语［M］. 北京：机械工业出版社，2021.
［7］ 赵雯，凌双. 新导向职业英语［M］. 北京：高等教育出版社，2021.
［8］ 王军. 自动化专业英语［M］. 3版. 重庆：重庆大学业出版社，2018.

Answer Keys

Unit One　Electrical Engineering

Section One　Warming Up

Group Work　A B C D

Listen.　Benjamin Franklin.

Scripts：

Sally：What are you doing, Yang Kang?
Yang Kang：Oh, I'm searching for some information about Benjamin Franklin.
Sally：Benjamin Franklin? What are his contributions?
Yang Kang：Benjamin Franklin made some important discoveries in electricity. And he is also famous for his inventions such as the lightning rod, bifocal glasses, Franklin stove etc.

Section Two　Dialogue：Am I Qualified to be an Electrician

对话译文如下。

我有资格当一名电工吗？

杨康：看这个招聘广告，电业局在招聘专业的有经验的电工。你觉得我能胜任吗？
莎莉：当然啦。你已经完成了电工培训和学徒期，并且获得了电工证。
杨康：现在似乎需要更多的电工了，工资也高了。
莎莉：因为电是现代社会如此重要的一部分。我们需要在不同的地点工作的电工，从商业、住宅物业、商业项目到建筑区。他们是我们社会的必要组成部分。
杨康：电气系统和部件的安装、装配、重新布线、修理、测试和维护均由电工负责。它可以涉及任何事情，从修理灯泡到维护太阳能电池板。我多希望对电气术语有更扎实的理解啊。
莎莉：由于技术总是领先一步，并且发展速度快，电工需要不断地学习。为了与时俱进，并了解所有新工艺是如何工作的，他们必须参加培训课程。
杨康：确实是这样，我非常同意。并且具备良好的观察技能将有助于识别电气故障，处理工作中遇到的各种问题并避免事故。

莎莉：现在你有信心得到这份工作了吗？

杨康：是的。

Find Information

Task Ⅰ.

1. Because he'd like to be an electrician in the Electric Power Bureau.

2. Not only electrical training and an apprenticeship, but also the senior electrician certificate.

3. It seems that more electricians are currently in demand and well-paid.

4. In a variety of locations, from businesses, residential properties, and commercial projects, to construction zones.

5. Yes, he is.

Task Ⅱ. 1–5 T T T F T

Cheer up Your Ears

1. professional   2. apprenticeship   3. certificate   4. vital   5. variety

6. rewiring   7. maintaining   8. constantly   9. observational   10. confident

Words Building

Task Ⅰ. 1. D   2. A   3. B   4. A   5. D

Task Ⅱ. 1. apprenticeship   2. electrician   3. commercials   4. current   5. maintenance

Task Ⅲ. 1–5 C E G J I   6–10 A H D B F

Section Three   Passage: Electrical Engineers

电气工程师

在当今的数字时代，电力真正维持着世界的运转，从基本的家庭维护到更复杂的交通灯、交通和技术系统，这些系统使我们的城市得以运转。

电气工程师是创造这些系统并使其平稳运行的设计师，他们致力于从国家电网到手机和智能手表内的微芯片的一切工作。电气工程师参与设计、开发、测试和指导电气设备的制造，如雷达和导航系统、电动机、发电机或通信系统。他们还负责设计航空和汽车系统。

作为一名电气工程师，你有机会影响世界的运行方式。以下是提高电气工程技能的一些步骤。首先，关注学业，熟悉专业知识。工程课程还经常教你所需的软技能，如沟通和合作。其次，参加尽可能多的培训课程将有助于提高你的技能；你还可以向你的同事和客户征求有关你工作的反馈意见，以确定需要改进的方面。再次，反馈可以帮助你识别这些方面，并决定未来如何提高。最后，你要获得电气工程师资格证。

Answer Keys

Find Information

Task Ⅰ.

1. Because electricity truly keeps the world running, from basics of maintaining our homes to the more complex systems of traffic lights, transportation and technology that keep our cities running.

2. Electrical engineers are the designers that create these systems and keep them running smoothly.

3. They are involved in designing, developing, testing and directing the manufacture of electrical equipment. They are also responsible for designing aviation and automotive systems.

4. They need to be familiar with expertise and soft skills.

5. Absolutely.

Task Ⅱ. 1-5 T T F T T

Cheer up Your Ears

1. maintaining   2. smoothly   3. smart watches   4. manufacture   5. responsible
6. impact   7. expertise   8. communication   9. enhance   10. certifications

Words Building

Task Ⅰ. 1. electrical equipment   2. traffic lights   3. cell phones and smart watches
4. electric motors   5. communication systems   6. 电网   7. 雷达和导航
8. 数字时代   9. 培训课程   10. 汽车系统

Task Ⅱ. 1. A   2. C   3. D   4. A   5. C

Task Ⅲ. 1. basically   2. transport   3. responsibility   4. involved   5. manufacturer

Task Ⅳ. 1. feedback   2. client   3. colleagues   4. expertise   5. power   6. certifications
7. generator   8. automotive

Task Ⅴ.

1. 在当今的数字时代，电力系统使我们的城市得以运转。
2. 电气工程师参与设计、开发、测试和指导电气设备的制造。
3. 电气工程师应该学好专业知识。
4. 此外，尽可能多地参加培训课程有助于提高技能。
5. 您还可以向您的同事和客户征求有关您工作的反馈意见，以确定需要改进的方面。

Listening

Task Ⅰ. 1. A   2. C   3. D   4. C   5. D

Scripts:

1. W: Hi, Jack. Did you watch the football game last night?
   M: Of course, I did. It was so exciting.
   Q: What are they talking about?

2. W: You're working with this bank now, aren't you?
   M: Yes, I've been working here for 2 years as an assistant manager.
   Q: What is the man's job position in the bank?

3. W: Sir, what would you like to order?
   M: Let me look at the menu for a moment first.
   Q: Where does this conversation most probably take place?

4. W: Good morning. What can I do for you?
   M: I'd like to change 200 U. S. dollars.
   Q: What did the man want to do?

5. M: Why are you interested in working with our company?
   W: I think I will have a better chance for career development here.
   Q: What is the man probably doing now?

Task Ⅱ. 1. our hearts   2. discussing   3. come true   4. possible   5. Without your efforts

Extensive Reading

电气系统设计

电气设计涉及电源转换、功率分配、照明系统、电信、文件报警系统、公共广播系统、闭路电视。要想擅长设计电气系统，您必须考虑以下几点。

1. 电子基础知识。

在开始设计电气系统之前，您需要了解电气组件、界面和基本原理。

2. 强调连接性。

我们生活在一个相互联系的世界。大多数设备都有几个天线。除了来自 WiFi 的高频信号，它们还可能包括可能导致潜在相间问题的传感器。因此，在设计电气系统时，必须包括 5G 网络的优化、物联网和专用校园网络。

3. 机械要求。

电气设计旨在承受一定水平的结构和操作要求。一些问题可能会对电气系统的整体设计造成挑战。

4. 系统的热条件。

当电气系统运行时，它往往会发热。您的设计必须使系统的热调节失效。必须进行热流分析。

5. 电路仿真。

电路仿真在原理图阶段起着至关重要的作用。在电路模拟过程中，您将能够在系统启动之前找到错误并解决问题。

6. 编码。

为了开发复杂的电气系统，设计工程师需要基于模型的开发软件。

7. 机电一体化。

今天，每一种产品都需要在所有可能的方式上变得智能。智能传感器和致动器是智能产品监测和控制的战略组件。此外，这些智能技术帮助用户提供智能系统所需的性能。

1. D   2. D   3. A   4. B   5. A

Unit Two   CAD & CAM

Section One   Warming Up

Group Work   C   E   A   D   B

Listen.   Make a presentation to the client and get feedback from him.

Scripts：

You should follow the CAD working procedure. First, receive a client and get information of his requirements for the new product. Second, refer to some books on designing and collect relevant materials about the other designs of the product. Third, work on a draft of the product and draw pictures on a computer.

Fouth, discuss with colleagues for new ideas and revise the draft to meet the client's demand. Fifth, make a presentation to the client and get feedback from him.

Section Two Dialogue：Introduction to CAD and CAM

对话译文如下。

CAD 和 CAM 简介

莎莉：你在用计算机做什么？

杨康：我正在学习如何使用 CAD 软件。CAD 是通过电子计算机技术应用于工程领域的产品设计的一种交叉技术。

莎莉：哦，我听说过 CAD。那么 CAD 和 CAM 有什么区别呀？

杨康：CAD 专注于产品或零件的外观和功能设计。CAM 专注于如何制作。每个工程进程都始于 CAD 世界。工程师们绘制 2D 或 3D 图纸，无论是汽车的曲轴、厨房水龙头的内部骨架，还是电路板中隐藏的电子元件。CAD/CAM 技术是许多人以生产自动化的名义几十年努力的结果。这是创新者和发明家、数学家和机械师的愿景，他们都在努力建设未来，用技术推动生产。

莎莉：哪些专业需要用到 CAD 软件呢？

杨康：所有行业的工程学科都使用 CAD 软件。如果你是设计师、绘图员、建筑师或工程师，就可能使用过 CAD 程序。

莎莉：有免费的 CAD 软件供初学者使用吗？

杨康：当然有了。

Find Information

Task Ⅰ.

1. He is learning how to use CAD software on the computer.

2. Yes. It is the cross technology of product design which is applied in engineering field by electronic computer technology.

3. CAD focuses on the design of a product or part, how it looks, how it functions. CAM focuses on how to make it.

4. 2D or 3D drawings. From a crankshaft for an automobile, the inner skeleton of a kitchen faucet, to the hidden electronics in a circuit board.

5. It is used by engineering disciplines across all industries.

Task Ⅱ. 1-5 T F T T F

Cheer up Your Ears

1. cross   2. engineering   3. functions   4. process   5. drawing

6. numerous   7. production   8. professions   9. disciplines   10. available

Words Building

Task Ⅰ. 1. A 2. D 3. A 4. D 5. C

Task Ⅱ. 1. applicants 2. professional 3. architecture 4. more numerous 5. innovate

Task Ⅲ. 1-5 C H G I A 6-10 E F J D B

Section Three　Passage：Application of CAD Technology

CAD 技术的应用

CAD 技术应用于许多不同的行业和职业，可以用于进行建筑设计及工业设计，生成装配和工程图纸以供制造。3D 打印、设计验证和确认都可以用 CAD 软件来实现。

在计算机辅助设计出现之前，设计需要用手工绘制。每个物体、直线或曲线都需要手工绘制。计算需要由工程师或设计师手动完成，这是一个非常耗时且容易出错的过程。

CAD 技术改变了这一切。CAD 与手工绘图相比具有多种优势，这使它现在绝对必不可少。设计可以在更少的时间内创建和编辑，也可以保存以备将来使用。CAD 图纸不限于一张纸的 2D 空间，可以从许多不同的角度查看，以确保适当的配合和设计。计算由计算机执行，从而更容易测试设计的可行性。设计可以实时共享和协作，大大减少了完成图形所需的总时间。

随着 CAD 程序变得越来越先进，它们将为设计师和工程师打开更多的可能性，并对越来越多的行业变得更加重要。

Answer Keys

Find Information

Task Ⅰ.

1. Yes, it has.

2. It's time-consuming and error-prone.

3. Designs can be created and edited in much less time, as well as saved for future use.

4. CAD drawings can ensure proper fit and design.

5. Yes. They will open up more possibilities to designers and engineers and become even more important to an ever-increasing range of industries.

Task Ⅱ. 1 – 5 T F F T F

Cheer up Your Ears

1. occupations   2. assembly   3. printing   4. Prior   5. object
6. Calculations   7. essential   8. created   9. viewed   10. ever-increasing

Words Building

Task Ⅰ. 1. application of CAD Technology   2. architectural design   3. create and edit
4. manual drawings   5. CAD drawings   6. 在……之前   7. 共享和协作
8. 越来越多的行业   9. 一个耗时的过程   10. 实时

Task Ⅱ. 1. A   2. D   3. C   4. A   5. B

Task Ⅲ. 1. applicant   2. Consumption   3. performances   4. advanced   5. is sent

Task Ⅳ. 1. assembly   2. objected   3. viability   4. architectural   5. advent   6. angle
7. verification   8. calculation

Task Ⅴ.

1. 在计算机辅助设计出现之前，设计需要用手工绘制。

2. CAD 与手工绘图相比具有多种优势，这使得它现在绝对必不可少。

3. CAD 图纸不限于一张纸的 2D 空间，可以从许多不同的角度查看，以确保适当的配合和设计。

4. 设计可以实时共享和协作，大大减少了完成图形所需的总时间。

5. 随着 CAD 程序变得越来越先进，它们将为设计师和工程师打开更多的可能性，并对越来越多的行业变得更加重要。

Listening

Task Ⅰ. 1. B   2. A   3. C   4. D   5. B

Scripts:

1. M: Do you know when the coach is coming?

W: I just called the tour guide and it'll arrive in a few minutes.

Q: What can we learn from the conversation?

2. W: Jack, what can I do for you with your presentation?

M: Please send me the sales data for the last quarter.

Q: What does the man ask for?

3. W: Hello, reception, what can I do for you?

M: I'll catch the early flight tomorrow morning, so please give me a wake-up call at 5:30.

Q: What does the man ask the woman to do?

4. M: Hi, Linda, would you like to join in an evening party tomorrow?

W: I'd like to, but I have to meet my clients at the airport.

Q: Why can't the woman attend the evening party?

5. M: Hello, I would like to change the flight I've booked for next Monday.

W: No problem. Please tell me your name and flight number.

Q: Why does the man make the phone call?

Task Ⅱ. 1. important part 2. experience 3. birthplace 4. pick up 5. enjoy a cup

Extensive Reading

CAD/CAM 技术的下一步是什么?

以下趋势可能向我们展示 CAD/CAM 的下一个飞跃技术将会出现。

人工智能:将 AI 融入设计软件设计任务的自动化,通过预测设计错误来增强质量控制,(通过机器学习)为无须人工输入即可创建独特设计铺平道路。

云协作:云技术允许 CAD/CAM 超越工作场所的单台计算机通过 SaaS 实现通用访问(软件即服务)模型。这意味着几个人可以一次完成同一个项目,同时跨部门和地区共享变得更加容易。

虚拟现实:VR 头盔和 VR 眼镜可用于利用复杂 CAD 软件提供逼真的可视化效果。例如,建筑师现在可以提供仅作为数字模型存在的建筑的"漫游"。

定制:软件提供商正在从"一刀切"的解决方案转向提供配置 CAD/CAM 以适应您的工作环境的选项,并只选择特定工作所需的工具。这可能是一种通过削减普通用户可能永远不需要的数十项功能来提供可负担性的方法。

1. D 2. A 3. A 4. B 5. D

Unit Three Automation

Section One Warming Up

Group Work 1-6 B F C E A D

Listen. He is visiting an automated factory.

Scripts: Sally: Good morning, Mr. Yang.

Yang Kang: Good morning, Ms. Wood.

Sally: Welcome to our factory.

Yang Kang: Thank you. I've been looking forward to visiting your factory.

Sally: I'll show you around and explain the operation as we go along.

Section Two Dialogue: Visiting an Automated Factory

参观自动化工厂

莎莉：早上好，杨先生。

杨康：早上好，伍德女士。

莎莉：欢迎到我们工厂来。

杨康：谢谢，我一直都盼望着参观贵厂。

莎莉：我带您四处看看，边走边给您讲解操作。

杨康：那太好了。工厂是全自动的吗？

莎莉：不完全是。我们的生产过程是部分自动化的。我们在生产线上使用机器人进行常规装配工作，但有些工作仍然是手工完成的。

杨康：那生产线的零件供应呢？

莎莉：嗯，零件是用条形码系统从储藏室自动挑选出来的。还有一个自动送料机，把它们送到生产线开始处的传送带上。

杨康：那小零件呢？

莎莉：它们由自动车辆——机器人卡车——运到工作站，这些车辆在工厂周围的导轨上运行。

杨康：那太好了。

莎莉：是的。我们现在休息一下好吗？

杨康：好的。

Find Information

Task I.

1. Sally.

2. Not completely. The production process is partially-automated.

3. Robots are used for routine assembly jobs on the production line.

4. The parts are automatically selected from the store room using a bar-code system. And there is an automatic feeder which takes them to the conveyor belt at the start of the production line.

5. They're transported to the workstations on automated vehicles—robot trucks—which run on guide rails around the factory.

Task II. 1—5 T F T F T

Cheer up Your Ears

1. visiting 2. operation 3. fully 4. partially 5. robots
6. supply 7. bar-code 8. feeder 9. components 10. workstations

Words Building

Task Ⅰ. 1. C 2. C 3. B 4. D 5. A

Task Ⅱ. 1. helpful 2. automation 3. operator(s) 4. operation 5. production

Task Ⅲ. 1–5 D F E I A 6–10 G J B H C

Section Three　Passage：Advantages and Disadvantages of Automation

自动化的利与弊

机械化的进一步发展以自动化为代表，这意味着利用控制系统和信息技术来减少生产商品对体力劳动和脑力劳动的需求。

过去的一个世纪，自动化对工业产生了巨大影响，世界经济从工业岗位转变为服务业岗位。

下面总结了自动化的主要优点和缺点。

1. 优点。

加快社会发展进程；

在重体力工作或单调工作方面取代人工操作员；

节省时间和资金，人工操作员可以从事更高水平工作；

在危险环境（火灾、太空、火山、核设施、水下）取代人工操作人员；

执行任务时有更高的可靠性和精确性；

改善经济，提高生产力。

2. 缺点。

对环境的灾难性影响（污染、交通、能源消耗）；

由于机器取代了人类，失业率急剧上升；

技术限制，目前的技术无法使所有所需的任务自动化；

作为自动化系统，安全威胁可能具有有限的智能水平，并可能出错；

初始成本高，因为新产品的自动化需要大量的初始投资。

自动化有利有弊，但是整体而言，自动化的迅速发展还是利远大于弊。

Find Information

Task Ⅰ.

1. No. It reduces the need for both physical and mental work to produce goods.

2. Because automation has changed the world economy from industrial jobs to service jobs.

3. In the fire, space, volcanoes, nuclear facilities, and underwater.

4. No, it isn't.

    5. Inform the supervisor immediately.

Task Ⅱ. 1 – 5 F F T F T

Cheer up Your Ears

    1. automation   2. control; reduce   3. industries   4. society   5. physical

    6. time   7. dangerous   8. environment   9. unemployment   10. advantages; disadvantages

Words Building

Task Ⅰ. 1. automated factory   2. control system   3. information technologies   4. physical work

    5. production line   6. 失业率   7. 脑力劳动   8. 人工操作员   9. 能源消耗

    10. 安全威胁

Task Ⅱ. 1. C   2. B   3. C   4. A   5. D

Task Ⅲ. 1. employee   2. to improve   3. development   4. polluted   5. industry

Task Ⅳ. 1. automated system   2. intelligence   3. pollution   4. speed up   5. physical

    6. Automation   7. production line   8. show...around

Task Ⅴ.

    1. 在上个世纪，自动化对工业产生了巨大影响，世界经济从工业岗位转变为服务业岗位。

    2. 在重体力工作或单调工作方面取代人工操作员。

    3. 节省时间和资金，人工操作员可以从事更高水平工作。

    4. 在危险环境（火灾、太空、火山、核设施、水下）取代人工操作人员。

    5. 自动化有利有弊，但是就整体而言，自动化的迅速发展还是利远大于弊。

Listening

Task Ⅰ. 1. C   2. D   3. C   4. A   5. B

Scripts：

    1. W：Mr. Smith, this is an agenda for tomorrow's meeting.

        M：Please tell all the managers to attend it.

        Q：What does the man ask the woman to do?

    2. W：Would you like to go to the lecture with me?

        M：I'd like to. But I have to finish my paper today.

        Q：What does the man mean?

    3. W：Can you tell me why you want to apply for this job?

        M：I am really attracted by the salary you offer.

        Q：Why does the man want to apply for this job?

4. W: The price at your product is about 15% higher than last year.
   M: Yes, because the labour cost has increased.
   Q: Why has the price increased?

5. W: George, do you have any suggestion to improve our product sales?
   M: Why not carry out a market survey among our customs?
   Q: What does the man suggest for improving sales?

Task Ⅱ. 1. success 2. wonderful 3. remarkable 4. great ideas 5. get in touch

Extensive Reading

传感器

几乎每一个工业自动化过程都需要使用传感器和换能器，这些传感器和换能器是非常先进的设备，能够测量和感测环境，并将物理信息（如光、压力、温度和位置的变化）转换为电信号。传感器拾取待测量的信息，换能器将其转换成电信号，该电信号可由系统的控制单元直接处理。

由于工业和科学测量的重要性，传感器被广泛应用于各种领域，如医学、工程、机器人、生物和制造业。传统机器很难测量产品尺寸的微小差异，因此传感器特别有用，因为它们可以精确到 0.000 13 mm。它们还可以检测温度、湿度和压力，获取数据并改变制造过程。传感器也是机器人等先进机器的重要组成部分。

传感器有两种类型：模拟传感器和数字传感器。模拟传感器使用测量的电压或量表示的数据进行操作，而数字传感器具有数字输出的功能，可以直接传输到计算机。

通常用于制造的传感器分为机械、电气、磁性和热传感器，但也可以被分类为声学、化学、光学和辐射传感器。此外，根据其传感方法，它们分为触觉或视觉传感器。触觉传感器对触摸、力或压力敏感，用于测量和记录接触表面与环境之间的相互作用。这些传感器用于日常物体中，如升降按钮和灯，它们通过触摸底座来打开和关闭。视觉传感器则以光学方式感知物体的存在、形状和运动。它们在监视系统、环境和灾害监测以及军事应用中变得越来越重要。

1-5 B A C B A

Unit Four Workplace Safety

Section One Warming Up

Group Work 1-6 D E C B F A

Listen. She advises him to provide safety training for the new employees.

Section Two Dialogue: Talking about Electrical Safety

谈论电气安全

莎莉：怎么了，杨康？

杨康：没电了。

莎莉：你检查过保险丝盒吗？

杨康：是的，保险丝烧断了。我换过保险丝，但是现在发动机仍然是断开的。

莎莉：可能线路有个地方松了造成安全开关跳闸。如果你自己不能搞定，可以找一个电气工程师，否则会很危险。

杨康：好的，谢谢。看来对我们来说了解一些车间用电知识很有必要啊。

莎莉：当然。特别是电气安全守则必须掌握。

杨康：你能给我一些提示吗？

莎莉：好的。工作之前，确保你接受过电气安全方面的培训。并且你必须时刻遵守安全指示。

杨康：明白。

莎莉：然后，确认所有可能造成危险的电源。永远不要假设设备或系统已经断电。记得在触碰前一定要测试。把你要操作的机器或其他设备上锁挂牌。维修前必须关闭电源。

杨康：听起来很有用。

莎莉：而且，选择合适的个人防护用品也很重要。

杨康：知道了。非常感谢。

Find Information

Task Ⅰ.

1. There's no power.

2. Yes, he has.

3. If Yang Kang can't fix it himself, call in an electrician.

4. Before you work, make sure you are trained in electrical safety. And you must follow the safety instructions at all times. (Then, identify all of the possible electric sources that could cause hazards. Never assume that the equipment or system is de-energized. Remember to test it before you touch it. Lock out/tag out machinery or other equipment you will work on. Turn off power before servicing. Choose the right personal protective equipment.)

5. PPE is short for personal protective equipment.

Task Ⅱ. 1–5 T F T F F

Cheer up Your Ears

1. electrician 2. electricity 3. rules 4. safety 5. instructions

169

6. hazards   7. system   8. Lock out   9. Turn off   10. protective

Words Building

Task Ⅰ.   1. C   2. B   3. D   4. B   5. A

Task Ⅱ.   1. electricity   2. dangerous   3. protection   4. training   5. to check

Task Ⅲ.   1-5 D G E H J   6-10 I A C F B

Section Three   Passage：Safety Rules in the Workshop

车间安全规则

必须注意安全问题，以确保工厂的安全生产。工人们必须意识到他们周围存在的危险和风险：三分之二的工业事故是由个人的粗心大意造成的。为了避免或减少事故，工作时必须同时采取保护和预防措施。

1. 一定要认真听从生产主管的指示，并遵从操作规程。如果主管没有给你安全指示，就不要使用机器。

2. 如有机器损坏，应立即向主管报告，以免引起事故。

3. 所有员工应穿工作服，佩戴工作证。此外，技术人员和维护人员应佩戴安全眼镜。

4. 时刻穿防护围裙，因为它可以保护你的衣服，并把宽松的衣服（如领带）固定住。

5. 必须穿防护鞋，可以预防或减少脚部伤害。车间里不允许穿着拖鞋、凉鞋或运动鞋。

6. 在车间里要有耐心，不要着急。

7. 不要在车间里奔跑，你可能会"撞"到机器或其他人，造成事故。

8. 不要把包袋带进车间，因为可能会把别人绊倒。

9. 知道车间里紧停按钮的位置。

10. 不舒服时不要操作机器。

Find Information

Task Ⅰ.

1. In order to avoid or reduce accidents.

2. Lest they cause an accident.

3. Because they can prevent or minimize foot injury.

4. Because you may "bump" into a machine or another person and cause an accident.

5. Because people can trip over bags.

Task Ⅱ.   1-5 F F F T F

Cheer up Your Ears

1. safety   2. dangers；carelessness   3. instructions   4. damage   5. safety glasses

6. foot   7. patient   8. Bags   9. stop   10. uncomfortable

Words Building

Task I. 1. safety rules   2. protective apron   3. work clothes   4. protective glasses   5. electric safety   6. 生产总监   7. 工业事故   8. 个人防护用品   9. 急停按钮   10. 防护和预防措施

Task II. 1. C   2. B   3. B   4. A   5. D

Task III. 1. prevention   2. protection   3. emergent   4. comfortable   5. carelessness

Task IV. 1. trip over   2. dangers   3. In addition to   4. In order to   5. industrial accident   6. operate   7. protective glasses   8. pay attention to

Task V.

1. 工人们必须意识到他们周围存在的危险和风险：三分之二的工业事故是由个人的粗心大意造成的。
2. 如有机器损坏，应立即向主管报告，以免引起事故。
3. 时刻穿防护围裙，因为它可以保护你的衣服，并把宽松的衣服（如领带）固定住。
4. 不要在车间里奔跑，你可能会"撞"到机器或其他人，造成事故。
5. 知道车间里紧停按钮的位置。

Listening

Task I. 1. C   2. D   3. A   4. B   5. D

Scripts：

1. W：What do you usually do in the morning?
   M：We go to the nearby lake and walk around it.
   Q：What does the man usually do in the morning?

2. W：Are you ready to order, Sir?
   M：Yes, I'd like fish and chips and a cup of coffee.
   Q：Where does the conversation most probably take place?

3. M：Could you deliver the fruits by train, please?
   W：Sorry, Sir. For fruits, we only deliver them by air.
   Q：What kind of goods does the man want to deliver?

4. M：My car broke down again, and it is in the repair shop now.
   W：If I were you, I would buy a new one.
   Q：What is the woman's advice?

5. W：Good evening, Beijing Restaurant. May I help you?
   M：I'd like to book a table for ten tomorrow evening.
   Q：What does the man want to do?

Task II. 1. relax   2. walk   3. strong   4. in the ocean   5. bus

Extensive Reading

消防安全

　　从学校、医院、超市到工作场所，所有公共建筑都需要消防安全计划。通常，建筑物的所有者负责编制这个计划。一旦它获得首席消防官员的批准，业主必须负责培训所有员工。

　　疏散演习是员工紧急疏散程序培训的一个非常重要的部分。所有建筑物应至少每年进行一次演习。应检查演习，记录完成疏散所需的时间，并且没有任何问题和不足。每次演习后，应召开一次会议，以评估演习的成功与否，并解决可能出现的任何问题。

　　万一发生火灾怎么办？

　　1. 如果你看到火或烟，不要惊慌。保持冷静，快速移动，但不要奔跑。

　　2. 提醒负责人并拨打正确的火警电话。派人与消防队员联络，告诉他们火灾发生在哪。如果他们自己找，可能会损失宝贵的时间。

　　3. 只有在安全的情况下，才能营救处于危险中的人。

　　4. 如果可行，关闭所有门窗以控制火势，以防蔓延。

　　5. 只有在经过培训且安全的情况下，才可以使用合适的消防设备灭火。

　　6. 遵照主管的指示，必要时准备撤离。

　　7. 条件允许的话，保存记录。

　　8. 撤离您所在的区域并检查所有房间，尤其是更衣室、卫生间、储藏室等。

　　9. 清点所有人数，并向主管报告下落不明的人员。

　　a □ 关闭所有门窗。

　　b □ 清点所有员工和访客的人数。

　　c □ 撤离您所在的区域并检查所有房间。

　　d □ 与消防员会面，并向他们提供火灾的详细信息。

　　e □ 保存记录。

　　f □ 准备撤离。

　　g □ 保持冷静，迅速行动。

　　h □ 向主管报告任何下落不明的人员。

　　i □ 拯救处于危险中的人。

　　j □ 拨打正确的火警号码。

　　k □ 尝试使用适当的消防设备灭火。

　　g-1，j-2，d-3，i-4，a-5，k-6，f-7，e-8，c-9，b-10，h-11

Unit Five　Autonomous Cars

Section One　Warming Up

Group Work C B A

Listen. American Wonder

Scripts:

Sally: Nowadays, automatic driving has become one of the selling points of cars, but in fact, it is not a new concept, and it has a long history of nearly a hundred years.

Yang Kang: Really? Then, when did people begin to study autonomous cars?

Sally: In August 1925, a wireless remote control vehicle named "American Wonder" was officially unveiled. Francis P. Houdina, an electric engineer of the United States Army, controlled the steering wheel, clutch, brake and other components remotely by means of wireless remote control.

Section Two Dialogue: How Far Away Are We from Autonomous Cars?

对话译文如下。

莎莉：现如今，自动驾驶已成为汽车的卖点之一，但实际上，自动驾驶汽车并非一个全新的概念，它已有近百年的悠久历史。

杨康：真的吗？那么人们是从什么时候开始研究自动驾驶的呢？

莎莉：1925年8月，一辆名为美国奇迹的无线遥控汽车正式亮相，该车由美国陆军电子工程师弗朗西斯 P. 霍迪尼通过无线电遥控的方式，来实现车辆方向盘、离合器、制动器等部件的远程操控。

杨康：真的令人难以想象啊。但它是真正意义上的无人驾驶汽车吗？

莎莉：不。1939年，通用汽车公司展出了世界上第一辆自动驾驶概念车——Futurama，这是一种由无线电控制的电磁场引导的电动汽车，该电磁场由嵌入道路的磁化金属尖刺产生。然而，直到1958年，通用汽车才将这一概念变为现实。

杨康：那么谷歌的自动驾驶汽车呢？是不是很有名？

莎莉：从2009年开始，谷歌就开始秘密开发无人驾驶汽车项目。在2014年，谷歌展示了没有方向盘、油门或刹车踏板的无人驾驶汽车的原型。

杨康：毋庸置疑，自动驾驶汽车的未来可期，不过在其发展的过程中，还面临着技术和伦理问题。我们所期望的高等级自动驾驶，还需要时日去发展和成熟。

Find Information

Task Ⅰ.

1. Nearly a hundred years.　2. In August 1925.　3. American Wonder.
4. Francis P. Houdina.　5. General Motors.

Task Ⅱ. 1–5 F T F F T

Cheer up Your Ears

1. automatic　2. history　3. wireless　4. engineer　5. displayed
6. electric　7. developing　8. demonstrated　9. promising　10. mature

Words Building

Task Ⅰ. 1. B   2. D   3. A   4. C   5. D

Task Ⅱ. 1. veil   2. official   3. magnetic   4. Automation   5. ethical

Task Ⅲ. 1-5 B C D G A   6-10 J I E F H

Section Three   Passage：Google's Self-driving Vehicles?

谷歌自动驾驶汽车

谷歌的自动驾驶汽车通过专门为帮助汽车准确感知周围环境而设计的传感器和处理接收到的信息的软件了解自己所在的位置和周围的事物。

1. 激光传感器。

该传感器使车辆能够360°了解其环境，从而了解车辆。可以同时感知自身前面、旁边和后面的物体。激光还可以帮助车辆确定其所处的位置。

2. 安全驾驶员。

驾驶员还每天测试车辆，给出如何使驾驶更安全和舒适的反馈报告。

3. 信息处理器。

软件对来自传感器的信息进行交叉检查和处理，以便准确地检测和区分车辆周围的不同物体，然后根据收到的所有信息做出安全驾驶决策。

4. 位置传感器。

该传感器位于轮毂中，用于检测车轮的旋转，以帮助车辆了解其所处的位置。

5. 方向传感器。

与人的内耳给人运动和平衡感的方式类似，这个位于汽车内部的传感器可以给汽车一个清晰的方向感。

6. 雷达。

该传感器检测前方很远的车辆并测量其速度，以便车辆能够安全减速或与道路上的其他车辆一起加速。

Find Information

Task Ⅰ.

1. Six.

2. Drivers also test the vehicles daily, reporting feedback on how to make the ride more safe and comfortable.

3. It detects vehicles ahead and measures their speed.

4. Orientation sensor.

5. In the wheel hub.

Task Ⅱ. 1-5 T F T T F

Cheer up Your Ears

1. Laser sensor   2. determine   3. sensors, software   4. feedback   5. processed
6. decisions   7. rotations   8. position   9. inner   10. sense

Words Building

Task Ⅰ. 1. perceive the surroundings   2. process the information   3. determine the. location
4. report feedback   5. safe and comfortable   6. 位置传感器   7. 方位传感器
8. 内耳   9. 运动感   10. 平衡感

Task Ⅱ. 1. A   2. A   3. C   4. C   5. A

Task Ⅲ. 1. difference   2. locate   3. orientation   4. determination   5. is based

Task Ⅳ. 1. surroundings   2. was based on   3. Slowing down   4. different   5. safety
6. sped by   7. is located   8. inner

Task Ⅴ.

1. 软件对来自传感器的信息进行交叉检查和处理，以便准确地检测和区分车辆周围的不同物体。
2. 该传感器位于轮毂中，用于检测车轮的旋转，以帮助车辆了解其所处的位置。
3. 车辆可以同时感知自身前面、旁边和后面的物体。
4. 方向传感器与人的内耳给人运动和平衡感的方式类似。
5. 这个位于汽车内部的传感器可以给汽车一个清晰的方向感。

Listening

Task Ⅰ. 1. A   2. D   3. C   4. A   5. B

Scripts：

1. M：Susan, do you know how long it takes to apply for a visa for China?
   W：5-7 work days, I'm afraid.
   Q：What are the two people talking about?

2. M：May I take your order, madam?
   W：Yes, I like a vegetable soup and Peking Duck, please.
   Q：Where does the conversation most probably take place?

3. W：I am calling to ask about the apartment you advertised in yesterday's newspaper.
   M：Yes, it's still available. Would you like to take a look?
   Q：Where does the woman get the information of the apartment?

4. W：When can you finish your project report, David?
   M：At least in two weeks. I have to check all the statistics again.
   Q：What does the man have to do with his report?

5. M：Do you think Mr. Brown is qualified for this position?

W: Yes, he would be a perfect choice.
Q: What does the woman mean?

Task Ⅱ. 1. celebrate   2. set up   3. take the opportunity   4. support   5. best service

Extensive Reading

自动驾驶汽车

第一批真正意义上的自动驾驶汽车终于问世了。

Waymo 最初是作为谷歌的自动驾驶汽车项目而诞生的，该公司宣布，已在亚利桑那州凤凰城的部分地区免费提供无人驾驶汽车，在紧急情况下，前排座位上没有人接手。该集团表示，参加 Waymo 试验的公众，将能够在"未来几个月"通过一个应用程序订购这些车辆。

可能是最具革命性的新技术之一，也是最具广告宣传的无人驾驶汽车之一，一直是大型汽车制造商和科技公司竞争的焦点。不过，尽管许多团体正在街头测试这项技术，让后备司机驾驶，但大多数人认为，实现完全自动的方法至少还需要两年时间。7 年前，谷歌首次展示了其无人驾驶技术的基本版本，震惊了汽车行业，随后该公司投资逾 10 亿美元用于自动驾驶汽车研究。

竞争对手承认它在技术上仍然处于领先地位，尽管怀疑论者质疑人工智能是否足够好以应对路上可能发生的许多不可预见的（意料之外的）事件。

Waymo 相信，它是第一家达到无人驾驶汽车领域 4 级标准的公司，这意味着它的汽车可以在经过仔细规划和测试的预设区域完全自动驾驶。Uber、通用汽车、Aptiv PLC、宝马和其他公司一直在进行测试，以达到 4 级，但所有这些公司仍然让一个人坐在驾驶座上。

1. C   2. A   3. C   4. D   5. D

Unit Six　High Speed Trains

Section One Warming Up

Group Work　B C A

Listen.　In 2005

Scripts:

Yang Kang: When and where did the first high speed trains appear?

　　Sally: The first high speed trains appeared as early as 1933 in Europe and the U. S.

Yang Kang: Is "Shinkansen" a kind of high speed trains?

　　Sally: Yes, in the mid-1960s, Japan introduced the world's first high volume high speed train that operated with a standard (4 ft) gauge. It was called the Shinkansen and officially opened in 1964.

Yang Kang: Then, do you know when the high speed trains first appeared in China?

Sally: In 2005, Beijing-Tianjin Intercity Railway.

Section Two Dialogue: History of High Speed Trains

对话译文如下。

杨康：莎莉，你在干什么？

莎莉：我在读一本关于高速列车的书。

杨康：什么样的火车可以被称为是高速列车呢？

莎莉：根据运行速度和所用技术，高速列车标准又有不同。在欧盟，运行时速至少125 mi 的为高速列车，而在美国，该标准为 90 mi。

杨康：第一辆高速列车在何时，何地出现的呢？

莎莉：最早的高速列车于1933年出现在欧洲和美国。

杨康："新干线"是一种高速列车吗？

莎莉：是的，二十世纪六十年代中期，日本推出了世界上首款标准轨距（4 ft）运行的大容量高速列车。列车名为"新干线"（Shinkansen），于1964年正式开通运行，往来于东京和大阪，时速约135 mi。这就是高速列车的早期发展史。

杨康：那么，你知道中国高速列车最早出现在什么时候吗？

莎莉：2005年开通的京津城际铁路。

Find Information

Task I.

1. In the European Union, high speed trains are those which travels 125 mi/h or faster.

2. In the United States, they are that travel 90 mi/h or faster.

3. The first high speed trains appeared as early as 1933 in Europe and the U. S.

4. Japan.

5. In 2005, Beijing-Tianjin Intercity Railway.

Task II. F F T T T

Cheer up Your Ears

1. appeared 2. provided 3. development 4. introduced 5. faster
6. travel 7. technology 8. called 9. high speed trains 10. Intercity

Words Building

Task I. 1. A 2. B 3. D 4. C 5. A

Task II. 1. introduction 2. developed 3. appearance 4. provides 5. constitution

Task III. 1 – 5 F E D C B 6 – 10 A G J I H

Section Three Passage：High Speed Trains in China

中国的高速铁路

日本曾因发达的高铁网络而闻名，早在1964年，"新干线"（或叫"子弹头列车"）就为世界熟知。尽管日本在铁路技术方面仍处于领先地位，但在高铁运营方面，中国才是当之无愧的世界第一。

大约20年前，中国第一批客运高铁投入运营；中国已经铺设超过4万千米的高铁轨道，总里程居世界第一位。2017年，中国的"复兴号"投入使用，它是世界上运行速度最快的高速列车，时速高达350 km，使北京到上海的时间缩短到4.5 h。

在中国，高铁是指时速在250～350 km之间的客运列车。除了个别省份，中国几乎所有的省份都开通了高铁。2017年7月，随着内蒙古第一条高速铁路通车，全国就只剩西藏和宁夏还未开通高铁。但按照计划，高铁网络还会继续扩张，相信这两地很快也会开通高铁。

在辽阔的中国土地上，高铁网络的发展为人们提供了高速且相对经济的出行方式。

Find Information

Task Ⅰ. 1. In 1964 2. Japan 3. China 4. Four and a half hours 5. Only Xizang and Ningxia.

Task Ⅱ. 1-5 F T T F T

Cheer up Your Ears

    1. vast distances 2. serviced 3. lack 4. defined 5. provinces

    6. reducing 7. launched 8. operation 9. constructed 10. technology

Words Building

Task Ⅰ. 1. reduce travel time 2. be famous for 3. be known as 4. be a leader in

    5. be serviced by 6. 辽阔的土地 7. 相对经济 8. 不断的扩大 9. 铁路技术

    10. 投入运营

Task Ⅱ. 1. B 2. C 3. D 4. A 5. D

Task Ⅲ. 1. technical 2. definition 3. expansion 4. operation 5. construction

Task Ⅳ. 1. the crown of 2. a leader in 3. makes for 4. be famous for 5. way back

    6. is famous as 7. is famous for 8. be put into operation

Task Ⅴ.

1. 尽管日本在铁路技术方面仍处于领先地位，但在高铁运营方面，中国才是当之无愧的世界第一。

2. 大约20年前，中国第一批客运高铁投入运营，中国已经铺设超过4万千米的高铁轨道，总里程居世界第一位。

3. 2017年，中国的"复兴号"投入使用，它是世界上运行速度最快的高速列车，时速高达 350 km。

4. 在中国，高铁是指时速在 250～350 km 的客运列车。

5. 在辽阔的中国土地上，高铁的发展为人们提供了高速且相对经济的出行方式。

Listening

Task Ⅰ. 1. A 2. C 3. B 4. D 5. A

Scripts：

1. W：Youth Travel Agency. What can I do for you, Sir?
   M：I've read your advertisement about the tour to China; I'm calling to inquire about it.
   Q：What does the man want to know?

2. M：I bought this washing machine last week, but it isn't working now.
   W：Have you brought the receipt with you?
   Q：What does the woman want to see?

3. M：Oh, I have left my smart phone in the office.
   W：Don't worry. Let's go back for it.
   Q：What will the two speakers probably do?

4. M：Marry, would you please work out a schedule for tomorrow's meeting?
   W：All right. I'll do it right away.
   Q：What will the woman probably do?

5. M：How was your winter vacation, Amy?
   W：Great, I spent the whole time working as a volunteer in the museum.
   Q：What do we know about the woman during the vacation?

Task Ⅱ. 1. high speed train 2. flights 3. compared with 4. friendly 5. additional

Extensive Reading

全球最快九大列车

列车固然不能像飞机一样飞越海洋，但这并不意味着其速度不能像飞机一样快。当今世界，有些列车的速度已经可以与飞机媲美。

第一名　中国上海磁悬浮列车，时速 267.8 mi。

上海磁悬浮列车是世界上速度最快的列车，磁悬浮是磁力悬浮的简称，指某一物体在磁场中悬浮或飘浮。

第二名　中国和谐号 CRH380A 型电力动车组，时速 236.12 mi。

中国铁路和谐号 CRH380A 型电力动车组是全球速度第二快的运营列车。

第三名　意大利 AGV Italo，时速 223.6 mi。

AVG Italo 是欧洲速度最快的列车。该列车在罗马和那不勒斯之间运行。

第四名 西班牙西门子 Velaro E 型列车 AVS 103，时速 217.4 mi。

Velaro E 是德国西门子公司开发的一款高速列车，在西班牙，该车型被称为 AVS 103。该列车在巴塞罗那和马德里之间运行。

第五名 西班牙 Talgo 350，时速 217.4 mi。

Talgo 350 高速列车由西班牙国营铁路公司运营，最高时速达 217.4 mi，在马德里和巴塞罗那之间运行。

第六名 日本新干线列车 E5 系"隼"号，时速 198.8 mi。

新干线 E5 系电力动车组是如今日本速度最快的高速列车，由东日本旅客铁路公司于 2011 年 3 月 5 日起在东京与青森县之间运营。

第七名 法国阿尔斯通 Euro-duplex，时速 198.8 mi。

Euroduplex 是第三代 TGV Duplex 高速列车，由法国国家铁路公司运营，连接法国、瑞士、德国和卢森堡铁路网络。

第八名 法国 TGV Duplex，时速 198.8 mi。

TGV Duplex 是法国速度最快的列车，由法国国家铁路公司于 2011 年 12 月开始运营。

第九名 意大利 ETR 500 型红箭特快，时速 186.4 mi。

ETR 500 型红箭特快是意大利速度最快的列车，由意大利铁路公司运营。该列车在米兰—罗马—那不勒斯铁路线上运行，每天 72 个车次。

1. C   2. A   3. B   4. D   5. A

Unit Seven Robots

Section One Warming Up

Group Work D A B C

Listen. Robot vacuum cleaner

Scripts：

Vacuuming can be a never-ending task, especially if you have pets.

Robotic vacuums have been making the chore easier for more than a decade now.

The Pros of Robot vacuum cleaner are as follows：

Scheduled cleaning；

Clean in the places where humans can't easily reach；

Spot cleaning；

Less noise；

Plugging itself into its charging station when running out of battery；

Identifying obstacles.

Section Two Dialogue：Do You Like Robots to Work at Your Home？

对话译文如下。

你喜欢机器人在家里工作吗?

杨康:莎莉,祝贺你得到这份工作。

莎莉:谢谢你。但我每天下班后都感觉筋疲力尽。我渴望下班后可以享受更轻松的生活。

杨康:那么你需要一个家用机器人。我非常喜欢机器人。你不认为他们能完全理解我们给他们的指示很神奇吗?

莎莉:你喜欢机器人在家里工作吗?

杨康:当然。我很满意机器人在我家里做我非常不喜欢的事情。我可以设置吸尘器来做家务。多功能机器人炊具能够油炸、蒸、烤、慢煮和执行任何其他操作。

莎莉:哇,听起来棒极了。大声连接的家用机器人已经成为你生活的一部分。他们是超级帮手,让生活更轻松。

杨康:重要的是机器人可以保证你家的安全。智能设备可以相互连接。机器人可以成为从咖啡机到锁门的一切事物的控制中心。您可以从世界任何地方访问它!

莎莉:是的,我可以用手机控制自动宠物喂食器。家用机器人是一个完美的伙伴,让我可以随时观察猫咪们的状况,而无须担心它们。

Find Information

Task Ⅰ.

1. Yes, it does.

2. She needs a household service robot.

3. Because he is pretty content with robots working at home doing things that he really doesn't like.

4. Smart devices can connect with each other. A robot can be the control center for everything from your coffee machine to locking doors. And you can access it from anywhere in the world!

5. Yes.

Task Ⅱ. 1-5 T T F T T

Cheer up Your Ears

1. relaxing   2. obsessed   3. content   4. chore   5. Multi-function

6. Loud-connected   7. essential   8. access   9. automatic   10. household

Words Building

Task Ⅰ. 1. C 2. A 3. B 4. C 5. A

Task Ⅱ. 1. relaxed   2. congratulations   3. helpful   4. more content   5. obsessed

Task Ⅲ. 1. H 2. D 3. J 4. B 5. F 6. E 7. I 8. G 9. C 10. A

Section Three　Passage：Industrial Robots

工业机器人

工业机器人是一种自动化控制的、可改编程序的、多功能的操作器。

工业机器人最常见的用途是从事简单和重复的工业任务，包括装配线流程、拣选和包装、焊接以及类似功能。它们提供可靠性、准确性和速度。工人经常暴露在恶劣或困难的环境下，如高度、过量灰尘、噪声或有毒气体。通过引入机器人，可以提高效率，保障工人的安全。在检测工业产品缺陷的情况下，传统的手动检查成本高昂，检查员可能会遗漏问题或犯错误，而使用机器视觉检测缺陷可以大大提高效率和准确性。此外，工业机器人在标准化、安全性和智能化方面比人有更大优势。

随着开发的进行，机器人有望取代更多的人类工作，实际上帮助补充了劳动力。机器人技术对于推动从传统制造向智能制造的转变尤为重要。机器人已经成为实现智能制造的核心设备，也是智能工厂的组成部分。机器人产业是一个国家技术实力和高端制造水平的重要标志。

Find Information

Task Ⅰ.

1. An industrial robot is an automatically controlled, reprogrammable, multipurpose manipulator.
2. Simple and repetitive industrial tasks.
3. Harsh or difficult conditions such as heights, excess dust, loud noise or toxic gases.
4. Traditional manual inspection is expensive and inspectors may miss problems or make mistakes, while using machine vision to detect defects can greatly improve efficiency and accuracy.
5. Robotics.

Task Ⅱ. 1–5 T F T F T

Cheer up Your Ears

1. repetitive　2. processes　3. accuracy　4. exposed　5. Efficiency
6. standardization　7. labor force　8. transformation　9. manufacturing　10. technological

Words Building

Task Ⅰ. 1. loud noise　2. smart factories　3. manual inspection　4. core equipment
5. replenish the labor force　6. 高端制造　7. 标准化、安全性和智能化
8. 装配线流程　9. 技术实力　10. 推动转型

Task Ⅱ. 1. C　2. D　3. A　4. A　5. B

Task Ⅲ. 1. automatically　2. more programmable　3. repetition　4. assemble　5. reliable

Task Ⅳ. 1. to replenish　2. intelligent　3. has been transformed　4. manufacturing
5. defects　6. Labor　7. excess　8. to weld

Answer Keys

Task V.

1. 工业机器人是一种自动化控制的、可改编程序的、多功能的操作器。
2. 工业机器人最常见的用途是用于简单和重复的工业任务,包括装配线流程、拣选和包装、焊接以及类似功能。
3. 工人经常暴露在恶劣或困难的环境下,如高度、过量灰尘、噪音或有毒气体。
4. 随着开发的进行,机器人有望取代更多的人类工作,并有效地帮助补充劳动力。
5. 机器人产业是一个国家技术实力和高端制造水平的重要标志。

Listening

Task Ⅰ. 1. D 2. B 3. C 4. A 5. A

Scripts:

1. M: Excuse me, may I use this printer?
   W: Sorry, it's out of order.
   Q: What does the woman mean?

2. M: Hi, Mary. I've got a new job. The salary is good.
   W: Really? Congratulations.
   Q: Why does the woman congratulate the man?

3. W: Can I help you, sir?
   M: Yes, I want to rent a car for one week.
   Q: What does the man want to do?

4. W: Good morning. What's the problem?
   M: I'd like to change this shirt for a larger size.
   Q: Why does the man want to change the shirt?

5. W: Hello, Sales Department.
   M: Hello, I'm John Smith from ABC Company. May I speak to your manager?
   Q: Whom does the man want to speak to?

Task Ⅱ. 1. aily tasks 2. tourist season 3. followed 4. permits 5. concerning

Extensive Reading

机器人管理

时至今日,管理人员一直在很大程度上忽视了机器人,把它们当作工程问题而不是管理问题。这种现象不能再继续下去了:机器人正变得功能强大而且无处不在。由于企业可以使用机器人这种新型员工,它们可能需要重新考虑自身的人力资源战略了。

第一个问题是如何管理机器人本身。阿西莫夫在1942年确立了基本原则:机器人不能伤害人类。这一原则已经通过近年来的技术改良得以加强:现在的机器人对于它们周围的事物更加敏感,还可以遵照指示避免袭击人类。但是五角大楼的计划使得这一切变得更为复杂:从本质上来说,它们所制造的很多机器人将成为杀人机器。

第二个问题是如何处理人与机器人的关系中人类这一方的问题。劳动者们总是担心新技术会抢走他们的饭碗。当机器以人类的面孔出现时，这种担忧变得尤为强烈。恰佩克那部给机器人取名字的小说中描绘了这样一个世界：起初，机器人带来了很多好处，最终，它们却导致了大量的失业和不满。

所以，企业必须努力使工人相信机器人有助于提高产量，而不只是吞噬职位的外来者。它们需要展现给员工们看：坐在他们身边的机器人更多的是他们的帮手，而不是威胁。奥迪在引进工业机器人方面一直做得特别成功，因为这家汽车制造商让员工去发现那些机器人可以改进工作的领域，然后将监管那些机器人的岗位提供给员工。企业还需要阐明，机器人有助于保留富有国家的生产岗位。德国之所以没有像英国一样丧失如此多的生产岗位，原因之一就是，在德国，每一万名工人所对应的机器人数量是英国的 5 倍。

1. H 2. F 3. G 4. B 5. I 6. D 7. J 8. A 9. E 10. C

Unit Eight A Smart City

Section One Warming Up

Group Work B A C

Listen. Helsinki

Scripts:

Sally: Many cities have made positive contributions to the construction of smart cities and the upgrading of urban intelligence.

Yang Kang: Which city is the most famous one?

Sally: In Helsinki, you can use a smart phone to input your current location and destination. The application can intelligently arrange your journey and provide you with different options of route, travel time and price.

Yang Kang: Really, it's amazing.

Sally: Yes, but there's something more interesting. Residents install sensors in the refrigerator at home to remind the expiration date of food through applications, and advise residents to use food more properly instead of throwing it away after the expiration date.

Section Two Dialogue: What Does a Smart City Look Like?

对话译文如下。

杨康和莎莉在参观中国国际智慧城市博览会。

莎莉：你愿意住在这样的一个城市里吗？那里的建筑物为你关灯，自动驾驶汽车可以找到最近的停车位，甚至垃圾桶都知道什么时候满了？

杨康：当然。但是智慧城市到底是什么样子的呢？

莎莉：智慧城市是指利用物联网技术对城市资产和资源进行有效管理的大都市或国际大都市。

杨康：用哪些特征来界定智慧城市？

莎莉：每个智慧城市都有自己独特的目标，但所有城市都认可物联网技术的重要性，明白新兴的物联网和云端能力提供了一个重要的机会，让我们更好地了解城市中心的各种运作细节。

杨康：城市如何变得有智慧？

莎莉：要使城市变得有智慧，关键在于如何利用物联网技术和位置感知。

杨康：任何城市都可以变得有智慧吗？

莎莉：只要搭建了关键性的配套基础设施，市民、市政官员和当地政府都倾力支持，公私企业与各级政府也协同合作，任何城市，无论规模大小，都能发展成为智慧城市。

Find Information

Task Ⅰ.

1. A smart city refers to a metropolitan or cosmopolitan city that utilizes Internet of Things technology to effectively manage the city's assets and resources.

2. Each smart city will have their own unique objective, but they all acknowledge the importance of IOT technology and understand that emerging IOT and cloud capabilities offer a meaningful opportunity to better understand the intimate workings of an urban center.

3. The key for a city to become smart lies in how it can leverage both IOT technology and location-awareness.

4. Yes. (Any city can become a smart city, regardless of size, as long as it lays the key supporting infrastructure and has the commitment of citizens, municipal officials and local government and collaboration between public and private companies with the different levels of government.)

5. Yes.

Task Ⅱ. 1–5 F T T F T

Cheer up Your Ears

1. buildings 2. parking space 3. rubbish bins 4. look like 5. refers to
6. define 7. unique 8. lies in 9. size 10. offer

Words Building

Task Ⅰ. 1. A 2. D 3. C 4. B 5. B

Task Ⅱ. 1. to utilize 2. regardless 3. infrastructure 4. collaboration 5. a characteristic

Task Ⅲ. 1–5 C B F D G 6–7 E A

Section Three Passage: Smart Cities and Normal Cities—What Are the Differences?

智慧城市与普通城市有何区别？

全球各地的城市都在考虑如何使用新技术，智慧城市因而成为当下的时髦热词。智慧城市不再是对未来天马行空的设想。这些城市真实存在，遍布全球，正探索将智能技术融入人们日常生活的种种新途径。让我们探究一下，智慧城市与普通城市有何不同？换言之，是什么让智慧城市有了智慧？

智慧城市的概念对不同的城市与国家来说各不相同，这取决于它们的发展水平。但简单来说，智慧城市是指城市利用信息通信技术改善民生。市政府利用信息通信技术提升城市运行效率、与公众分享信息、提升政府服务质量及市民福利水平。

智慧城市的主要目的是通过人工智能有效利用城市基础设施，实现社会有效、高效运作。此外，在运用智能科技与数据分析改善市民生活质量的同时，致力于优化城市功能、推动经济增长。

Find Information

Task Ⅰ.

1. Smart cities.

2. Yes.

3. The main purpose of smart city is to create a society which can perform effectively and efficiently making effective use of city infrastructures through artificial intelligence.

4. A smart city is an urban area that uses information and communication technologies to ease up the livelihood of its people.

5. It focuses to optimize city functions and drive economic growth while improving quality of life for its citizens using smart technology and data analysis.

Task Ⅱ. 1–5 T F F T T

Cheer up Your Ears

1. buzzword 2. no longer 3. everyday lives 4. vary from 5. ease up
6. welfare 7. society 8. economic growth 9. close look 10. augment

Words Building

Task Ⅰ. 1. artificial intelligence 2. futuristic scenarios 3. everyday lives
4. the level of development 5. citizen welfare 6. 优化城市功能 7. 提升质量
8. 运行效率 9. 改善民生 10. 城市基础设施

Task Ⅱ. 1. A 2. B 3. D 4. A 5. A

Task Ⅲ. 1. simply 2. use 3. artificial 4. optimization 5. creative

Task Ⅳ. 1. dream up 2. ease up 3. depends on 4. in other words 5. share with

6. vary from   7. Simply put   8. make good use of

Task V.

1. 智慧城市的概念对不同的城市与国家来说各不相同，这取决于它们的发展水平。

2. 但简单来说，智慧城市是指城市利用信息通信技术来改善民生。

3. 市政府利用信息通信技术提升城市运行效率、与公众分享信息、提升政府服务质量及市民福利水平。

4. 智慧城市的主要目的是通过人工智能有效利用城市基础设施，实现社会有效、高效运作。

5. 此外，在运用智能科技与数据分析改善市民生活质量的同时，致力于优化城市功能、推动经济增长。

Listening

Task Ⅰ. 1. A   2. C   3. B   4. B   5. D

Scripts：

1. W：George, do you have any idea to improve our brand image?

   M：Oh, I've just talked about the brand image with Susan.

   Q：What are the two speakers talking about?

2. M：Hello, Customer Service. How can I help you?

   W：Yes, I'd like some information on your telephone banking service.

   Q：What is the woman asking about?

3. M：Excuse me, how can I get to the manager's office?

   W：Take the lift to the fifth floor. It's the third office on the left.

   Q：Where is the manager's office?

4. M：Would you help me to move into my new house this Saturday?

   W：I am really sorry. I have an appointment with my doctor.

   Q：What does the woman mean?

5. W：Could you go to the airport to meet Mr. Smith, the new engineer?

   M：Certainly. What's the flight number?

   Q：What does the woman ask the man to do?

Task Ⅱ. 1. inviting   2. boarding pass   3. gate   4. completed   5. repeat

Extensive Reading

智能城市不再是未来的一个梦

有了智能系统和新时代的公交网络，大城市的生活将变得更加幸福和高效。毕竟，根据联合国的一份报告，到2050年，预计将有超过60%的世界人口居住在城市中。使这些城市更适合许多人居住的答案在于创建"智能"城市。

这些城市将使用5G网络和"物联网"，使日常生活更安全，更便捷。波士顿，巴尔的摩、阿姆斯特丹和哥本哈根等城市已经在使用智能技术来改善居民的生活。但是，智慧城市到底能做什么？让我们看几个例子。

在美国的波士顿市和巴尔的摩市，智能垃圾桶可以感应到它们已经满了，并在需要清空时通知清洁工人。在荷兰的阿姆斯特丹，根据从城市周围的传感器（传感器）收集的实时数据，监视和调整交通流量和能源使用情况。在丹麦的哥本哈根，智能自行车系统使骑手在骑车时可以检查空气质量和交通拥堵情况。

智慧城市将具有互动性，使居民能够感觉到自己正在真正塑造自己的环境，而不仅仅是生活在其中。电信公司 Verizon 的智慧城市倡议负责人 Mrinalini Ingram 表示："拥有智慧城市的最重要原因之一是我们实际上可以以前所未有的方式与环境进行沟通。"

智慧城市还将使我们节省资源。通过使用传感器和 5G 网络监控水，煤气和电的使用，城市管理者可以弄清楚如何分配和更有效地节省这些资源。智慧城市也可以更紧密地监测二氧化碳和其他空气污染物的排放。

当然，将我们目前的城市变成未来的智慧城市将需要时间和金钱。但是，正如我们已经看到的那样，世界上越来越多的城市已经在以小规模的方式采用智能技术。例如，中国正在对上海和广州等大城市进行投资，以使其"更智能"。不久之后，甚至会有更多的城市开始开发自己的智能基础设施。

1. C 2. D 3. C

Listening Scripts

Unit One Electrical Engineering

Section One   Warming Up

Listen to the dialogue and find out which person above they are talking about.

Sally: What are you doing, Yang Kang?

Yang Kang: Oh, I'm searching for some information about Benjamin Franklin.

Sally: Benjamin Franklin? What are his contributions?

Yang Kang: Benjamin Franklin made some important discoveries in electricity. And he is also famous for his inventions such as the lightning rod, bifocal glasses, Franklin stove etc.

Section Two   Dialogue: Am I Qualified to Be an Electrician?

Listen and role-play the following dialogue.

Yang Kang: Look at the job advertisement. The Electric Power Bureau is looking for a professional and experienced electrician. Do you think I am qualified for it?

Sally: Absolutely. You have completed the electrical training and an apprenticeship. And also, you've obtained the senior electrician certificate.

Yang Kang: It seems that more electricians are currently in demand and well-paid.

Sally: Because electricity is such a vital part of modern-day society. We need electricians in a variety of locations, from businesses, residential properties, and commercial projects, to construction zones. They are necessary parts of our society.

Yang Kang: Installation, fitting, rewiring, repairing, testing, and maintaining electrical systems and components are all tasks an electrician is responsible for. It can involve anything from fixing a light bulb to maintaining a solar panel. I wish I had a more solid understanding of electrical terms.

Sally: Since technology is constantly one step ahead and moving at a rapid pace, electricians are perpetual students. To stay in touch with the growth, and to understand how all the new processes work, they have to take training courses.

Yang Kang: Exactly! I couldn't agree more. Besides that, having good observational skills will be helpful to identify electrical problems, deal with various problems and avoid accidents.

Sally: Are you confident of getting the job now?

Yang Kang: Yes.

Cheer up Your Ears

Listen and write down what you've heard. Then read and recite till you can use them fluently

1. The Electric Power Bureau is looking for a professional electrician.
2. Yang Kang has completed the electrical training and an apprenticeship.
3. He obtained the senior electrician certificate.
4. Electricity is such a vital part of modern-day society.
5. We need electricians in a variety of locations.
6. An electrician is responsible for installation, fitting, rewiring, repairing, testing, etc.
7. It can involve anything from fixing a light bulb to maintaining a solar panel.
8. Technology is constantly one step ahead and moving at a rapid pace.
9. Having good observational skills will be helpful to avoid accidents.
10. Are you confident of getting the job now?

Section Three　Passage: Electrical Engineers

In today's digital age, electricity truly keeps the world running, from basics of maintaining our homes to the more complex systems of traffic lights, transportation and technology that keep our cities running.

Electrical engineers are the designers that create these systems and keep them running smoothly, working on everything from the nation's power grid to the microchips inside our cell phones and smart watches. Electrical engineers are involved in designing, developing, testing and directing the manufacture of electrical equipment, such as radar and navigation systems, electric motors, generators, or communication systems. They are also responsible for designing aviation and automotive systems.

As an electrical engineer, you have the opportunity to impact the way that the world works. Here are some steps you can take to improve your electrical engineering skills. First of all, focus on schoolwork. Be familiar with expertise. Engineering courses also often teach you the soft skills you'll need, such as communication and cooperation. Furthermore, it can be helpful to attend training courses as much as possible to enhance your skills. Additionally, you can also ask your colleagues and clients for feedback on your work to determine areas that need improvement. Feedback

Listening Scripts

can help you identify those areas and decide how you'll improve in the future. Finally, pursue certifications available for electrical engineers.

Cheer up Your Ears

> Listen and write down what you've heard. Then read and recite till you can use them fluently

1. Electricity truly keeps the world running, from basics of <u>maintaining</u> our homes to the more complex systems that keep our cities running.

2. Electrical engineers create these systems and keep them running <u>smoothly</u>.

3. They work on everything from the nation's power grid to the microchips inside our cell phones and <u>smart watches</u>.

4. Electrical engineers are involved in designing, developing, testing and directing the <u>manufacture</u> of electrical equipment.

5. They are also <u>responsible</u> for designing aviation and automotive systems.

6. Electrical engineers have the opportunity to <u>impact</u> the way that the world works.

7. Focus on schoolwork. Be familiar with <u>expertise</u>.

8. Engineering courses also often teach you the soft skills you'll need, such as <u>communication</u> and cooperation.

9. It can be helpful to attend training courses to <u>enhance</u> your skills.

10. Pursue <u>certifications</u> available for electrical engineers.

Listening

> Task Ⅰ. Listen to five short dialogues and choose the best answer

1. W: Hi, Jack. Did you watch the football game last night?
   M: Of course, I did. It was so exciting.
   Q: What are they talking about?

2. W: You're working with this bank now, aren't you?
   M: Yes, I've been working here for 2 years as an assistant manager.
   Q: What is the man's job position in the bank?

3. W: Sir, what would you like to order?
   M: Let me look at the menu for a moment first.
   Q: Where does this conversation most probably take place?

4. W: Good morning. What can I do for you?
   M: I'd like to change 200 U.S. dollars.
   Q: What did the man want to do?

5. M: Why are you interested in working with our company?
   W: I think I will have a better chance for career development here.
   Q: What is the man probably doing now?

Task Ⅱ. Listen to the passage and fill in the blanks with the missing words or phrases

How great it is to see so many of you come and join us in celebrating the 15th anniversary of our travel magazine. From the bottom of <u>our hearts</u>, we thank you for being here. A little more than fifteen years ago, we were sitting at our regular jobs, <u>discussing</u> how we saw our future, when we came up with the idea of joining our two hobbies, traveling and writing. We never imagined that our tiny dream would <u>come true</u> so soon. There were many special people who joined us and made it <u>possible</u> to create the name that we have today. To all those people and those who joined us in our journey, I should say thank you again. <u>Without your efforts</u> we would never have been here.

Unit Two CAD & CAM

Section One   Warming Up

Listen to the following material and find out the last step in the CAD working procedure.

You should follow the CAD working procedure. First, receive a client and get information of his requirements for the new product. Second, refer to some books on designing and collect relevant materials about the other designs of the product. Third, work on a draft of the product and draw pictures on a computer. Fouth, discuss with colleagues for new ideas and revise the draft to meet the client's demand. Fifth, make a presentation to the client and get feedback from him.

Section Two   Dialogue: Introduction to CAD and CAM

Listen and role-play the following dialogue.

Sally: What are you doing on the computer?

Yang Kang: I'm learning how to use the CAD software. Computer aided design is the cross technology of product design which is applied in engineering field by electronic computer technology.

Sally: Oh, I've heard about CAD. And what's the difference between CAD and CAM?

Yang Kang: CAD focuses on the design of a product or part, how it looks and how it functions. CAM focuses on how to make it. The start of every engineering process begins in the world of CAD. Engineers make either a 2D or 3D drawing, whether that's a crankshaft for an automobile, the inner skeleton of a kitchen faucet, or the hidden electronics in a circuit board. CAD/CAM technology is the result of decades of efforts by numerous people in the name of production automation. It is the vision of innovators, inventors, mathematicians and machinists, who are all working to build the future and drive production with technology.

Sally: In which professions is CAD software needed?

Yang Kang: CAD software is used by engineering disciplines across all industries. If you're a designer, drafter, architect, or engineer, you've probably used CAD programs.

Sally: Is there any free CAD software available for beginners?

Yang Kang: Certainly.

Cheer up Your Ears

Listen and write down what you've heard. Then read and recite till you can use them fluently.

1. Computer aided design is a kind of <u>cross</u> technology.
2. CAD is normally used in <u>engineering</u> design.
3. CAD focuses on the design of a product or part, how it looks and how it <u>functions</u>.
4. The start of every engineering <u>process</u> begins in the world of CAD.
5. Engineers make either a 2D or 3D <u>drawing</u>.
6. CAD/CAM technology is the result of efforts by <u>numerous</u> people.
7. Innovators, inventors, mathematicians and machinists are all working to drive <u>production</u> with technology.
8. CAD technology can be applied in many <u>professions</u>.
9. CAD software is used by engineering <u>disciplines</u> across all industries.
10. There is some free CAD software <u>available</u> for beginners.

Section Three   Passage: Application of CAD Technology

CAD technology is used across many different industries and occupations, and can be used to make architectural designs and industrial designs, generate assembly and engineering drawings for manufacturing. 3D printing, verification and validation of design are possible in CAD software.

Prior to the advent of computer aided design, designs needed to be manually drawn. Every object, line or curve needed to be drawn by hand. Calculations needed to be done manually by an engineer or designer, a very time-consuming and error-prone process.

CAD technology changed all of this. CAD has a variety of advantages over manual drawings, which has made it absolutely essential nowadays. Designs can be created and edited in much less time, as well as saved for future use. CAD drawings are not limited to the 2D space of a piece of paper, and can be viewed from many different angles to ensure proper fit and design. Calculations are performed by the computer, making it much easier to test the viability of designs. Designs can be shared and collaborated on in real time, greatly decreasing the overall time needed to complete a drawing.

As CAD programs become more advanced, they will open up more possibilities to designers and engineers and become even more important to an ever-increasing range of industries.

Cheer up Your Ears

Listen and write down what you've heard. Then read and recite till you can use them fluently.

1. CAD technology is used across many different industries and occupations.
2. CAD can be used to generate assembly and engineering drawings for manufacturing.
3. 3D printing, verification and validation of design are possible in CAD software.
4. Prior to the advent of computer aided design, designs needed to be manually drawn.
5. Every object, line or curve needed to be drawn by hand.
6. Calculations needed to be done manually by an engineer or designer, a very time-consuming and error-prone process.
7. CAD has a variety of advantages over manual drawings, which has made it absolutely essential nowadays.
8. Designs can be created and edited in much less time, as well as saved for future use.
9. CAD drawings can be viewed from many different angles to ensure proper fit and design.
10. As CAD programs become more advanced, they will become more important to an ever-increasing range of industries.

Listening

Task Ⅰ. Listen to five short dialogues and choose the best answer.

1. M: Do you know when the coach is coming?
   W: I just called the tour guide and it'll arrive in a few minutes.
   Q: What can we learn from the conversation?

2. W: Jack, what can I do for you with your presentation?
   M: Please send me the sales data for the last quarter.
   Q: What does the man ask for?

3. W: Hello, reception, what can I do for you?
   M: I'll catch the early flight tomorrow morning, S please give me a wake-up call at 5:30.
   Q: What does the man ask the woman to do?

4. M: Hi, Linda, would you like to join in an evening party tomorrow?
   W: I'd like to, but I have to meet my clients at the airport.
   Q: Why can't the woman attend the evening party?

5. M: Hello, I would like to change the flight I've booked for next Monday.
   W: No problem. Please tell me your name and flight number.
   Q: Why does the man make the phone call?

Task Ⅱ. Listen to the passage and fill in the blanks with the missing words or phrases.

Good morning, everybody. Today, I'd like to introduce you to our tour for tea lovers. As you know, tea is an important part of Chinese tradition. You may have no idea about how the tea grows and how it is made. Our tour will enable you to experience the tea culture in China. Hangzhou is the birthplace of Longjing tea, which is one of the most famous green teas in China. During this tour, you will have the chance to go to a tea farm, pick tea leaves, visit a tea farmer's house, and learn the art of tea-making and enjoy a cup of Longjing tea. I hope you're pleased to travel with us to learn more about Chinese tea culture.

Unit Three Automation

Section One Warming Up

Listen to the dialogue and find out what Yang Kang is doing now.

Sally: Good morning, Mr. Yang.
Yang Kang: Good morning, Ms. Wood.
Sally: Welcome to our factory.
Yang Kang: Thank you. I've been looking forward to visiting your factory.
Sally: I'll show you around and explain the operation as we go along.

Section Two Dialogue: Visiting an Automated Factory

Listen and role-play the following dialogue.

Sally: Good morning, Mr. Yang.
Yang Kang: Good morning, Ms. Wood.
Sally: Welcome to our factory.
Yang Kang: Thank you. I've been looking forward to visiting your factory.
Sally: I'll show you around and explain the operation as we go along.
Yang Kang: That'll be most helpful. Is the factory fully-automated?
Sally: Not completely. Our production process is partially-automated. We use robots on the production line for routine assembly jobs but some of the work is still done manually.
Yang Kang: What about supply of parts to the production line?
Sally: Well, the parts are automatically selected from the store room using a barcode system. And there is an automatic feeder which takes them to the conveyor belt at the start of the production line.
Yang Kang: What about the smaller components?
Sally: They're transported to the workstations on automated vehicles—robot trucks—which run on guide rails around the factory.
Yang Kang: That's wonderful.
Sally: Yes. Shall we take a break now?

Yang Kang: Okay.

Cheer up Your Ears

Listen and write down what you've heard. Then read and recite till you can use them fluently.

1. I've been looking forward to visiting your factory.

2. I'll show you around and explain the operation as we go along.

3. Is the factory fully-automated?

4. Our production process is partially-automated.

5. We use robots on the production line for routine assembly jobs but some of the work is still done manually.

6. What about supply of parts to the production line?

7. The parts are automatically selected from the store room using a bar-code system.

8. There is an automatic feeder which takes them to the conveyor belt at the start of the production line.

9. What about the smaller components?

10. They're transported to the workstations on automated vehicles—robot trucks—which run on guide rails around the factory.

Section Three Passage: Advantages and Disadvantages of Automation

A further development of mechanization is represented by automation, which means the use of control system and information technologies to reduce the need for both physical and mental work to produce goods.

Automation has had a great impact on industries over the last century, changing the world economy from industrial jobs to service jobs.

The following sum up the main advantages and disadvantages of automation.

Advantages:

Speeding up the development process of society;

Replacing human operators in tasks that include hard physical or monotonous work;

Saving time and money as human operators can be employed in higher-level work;

Replacing human operators in tasks done in dangerous environments (fire, space, volcanoes, nuclear facilities, underwater);

Higher reliability and precision in performing tasks;

Economy improvement and higher productivity.

Disadvantages:

Disastrous effects on the environment (pollution, traffic, energy consumption);

Sharp increase in unemployment rate due to machines replacing human being;

Technical limitations as current technology is unable to automate all the desired tasks;

Listening Scripts

Safety threats as an automated system may have a limited level of intelligence and can make errors;

High initial costs as the automation of a new product requires a large initial investment.

Automation has both advantages and disadvantages, but as a whole, the advantages of the rapid development of automation still far outweigh the disadvantages.

Cheer up Your Ears

Listen and write down what you've heard. Then read and recite till you can use them fluently.

1. A further development of mechanization is represented by <u>automation</u>.
2. Automation means the use of <u>control</u> system and information technologies to <u>reduce</u> the need for both physical and mental work to produce goods.
3. Automation has had a great impact on <u>industries</u> over the last century, changing the world economy from industrial jobs to service jobs.
4. Speeding up the development process of <u>society</u>.
5. Replacing human operators in tasks that include hard <u>physical</u> or monotonous work.
6. Saving <u>time</u> and money as human operators can be employed in higher-level work.
7. Replacing human operators in tasks done in <u>dangerous</u> environments.
8. Disastrous effects on the <u>environment</u>, including pollution, traffic, energy consumption.
9. Sharp increase in <u>unemployment</u> rate due to machines replacing human being.
10. As a whole, the <u>advantages</u> of the rapid development of automation still far outweigh the <u>disadvantages</u>.

Listening

Task I. Listen to 5 short dialogues and choose the best answer.

1. W: Mr. Smith, this is an agenda for tomorrow's meeting.
   M: Please tell all the managers to attend it.
   Q: What does the man ask the woman to do?
2. W: Would you like to go to the lecture with me?
   M: I'd like to. But I have to finish my paper today.
   Q: What does the man mean?
3. W: Can you tell me why you want to apply for this job?
   M: I am really attracted by the salary you offer.
   Q: Why does the man want to apply for this job?
4. W: The price at your product is about 15% higher than last year.
   M: Yes, because the labour cost has increased.
   Q: Why has the price increased?
5. W: George, do you have any suggestion to improve our product sales?

M: Why not carry out a market survey among our customs?

Q: What does the man suggest for improving sales?

Task Ⅱ. Listen to the passage and fill in the blanks with the missing words or phrases.

We are now about to close the marketing conference. We owe thanks to every member of staff who made the conference a <u>success</u>. We would like to thank all of the speakers who have made <u>wonderful</u> and impressive speeches. Our thanks also go to every one of you for your contributions and your appreciation. The energy and the enthusiasm surrounding this conference have been <u>remarkable</u>. And now we'll go back to our business and put those <u>great ideas</u> into action. It's time to get back to start working on the next phase. All customers are out there waiting for us to <u>get in touch</u>.

Unit Four   Workplace Safety

Section One   Warming up

Listen to the dialogue and find out what Sally advises Yang Kang to do for the new employees.

Yang Kang: Regarding the weekly safety meeting, please let me know if any of you have any suggestions.

Sally: If you don't mind, I have a request.

Yang Kang: Please.

Sally: Would you mind providing safety training for the new employees?

Yang Kang: No problem.

Sally: It's very important for them to know about safety rules.

Yang Kang: Yes, of course.

Section Two   Dialogue: Talking about Electrical Safety

Listen and role-play the following dialogue.

Sally: What's wrong, Yang Kang?

Yang Kang: There's no power.

Sally: Have you checked the fuse box?

Yang Kang: Yes, the fuse had blown and I've changed it, but now the motor keeps cutting off.

Sally: There might be a loose connection somewhere that's making the safety switch trip. If you can't fix it yourself, call in an electrician, or it could be dangerous.

Yang Kang: Okay, thank you. It seems that it's necessary for us to know some electricity knowledge in the workshop.

Sally: Definitely! Especially the safety rules of electricity must be grasped.

Yang Kang: Could you give me some tips?

Sally: All right. Before you work, make sure you are trained in electrical safety. And you must follow the safety instructions at all times.

Yang Kang: I see.

Sally: Then identify all of the possible electric sources that could cause hazards. Never assume that the equipment or system is de-energized. Remember to test it before you touch it. Lock out/tag out machinery or other equipment you will work on. Turn off power before servicing.

Yang Kang: That sounds very useful.

Sally: It is also important to choose the right personal protective equipment (PPE).

Yang Kang: I got it. Thank you very much.

Cheer up Your Ears

Listen and write down what you've heard. Then read and recite till you can use them fluently.

1. If you can't fix it yourself, call in an <u>electrician</u>, or it could be dangerous.
2. It seems that it's necessary for us to know some <u>electricity</u> knowledge in the workshop.
3. Especially the safety <u>rules</u> of electricity must be grasped.
4. Before you work, make sure you are trained in electrical <u>safety</u>.
5. You must follow the safety <u>instructions</u> at all times.
6. Then identify all of the possible electric sources that could cause <u>hazards</u>.
7. Never assume that the equipment or <u>system</u> is de-energized.
8. <u>Lock out</u>/tag out machinery or other equipment you will work on.
9. <u>Turn off</u> power before servicing.
10. It is also important to choose the right personal <u>protective</u> equipment.

Section Three Passage: Safety Rules in the Workshop

Attention must be paid to safety in order to ensure a safe working practice in factories. Workers must be aware of the dangers and risks that exist all around them: two out of every three industrial accidents are caused by individual carelessness. In order to avoid or reduce accidents, both protective and precautionary measures must be followed while working.

1. Always listen carefully to the production supervisor and follow instructions. Do not use a machine if the supervisor has not shown you safely.

2. Report any damage to the machines to the supervisor immediately lest they cause an accident.

3. All staff shall wear work clothes and badge. In addition, technicians and maintenance personnel shall wear safety glasses.

4. Always wear a protective apron as it will protect your clothes and hold loose clothing such as ties in place.

5. Must wear protective shoes, which can prevent or minimize foot injury. No slippers, sandals or sneakers allowed in the workshop.

6. Always be patient and never rush in the workshop.

7. Do not run in the workshop, you could "bump" into a machine or another person and cause an accident.

8. Bags should not be brought into the workshop as people can trip over them.

9. Know where the emergency stop buttons are in the workshop.

10. Do not operate a machine if you feel uncomfortable.

Cheer up Your Ears

> Listen and write down what you've heard. Then read and recite till you can use them fluently.

1. Attention must be paid to safety in order to ensure a safe working practice in factories.

2. Workers must be aware of the dangers and risks that exist all around them: two out of every three industrial accidents are caused by individual carelessness.

3. Always listen carefully to the production supervisor and follow instructions.

4. Report any damage to the machines to the supervisor immediately lest they cause an accident.

5. In addition, technicians and maintenance personnel shall wear safety glasses.

6. Must wear protective shoes, which can prevent or minimize foot injury.

7. Always be patient and never rush in the workshop.

8. Bags should not be brought into the workshop as people can trip over them.

9. Know where the emergency stop buttons are in the workshop.

10. Do not operate a machine if you feel uncomfortable.

Listening

> Task I. Listen to 5 short dialogues and choose the best answer.

1. W: What do you usually do in the morning?
   M: We go to the nearby lake and walk around it.
   Q: What does the man usually do in the morning?

2. W: Are you ready to order, Sir?
   M: Yes, I'd like fish and chips and a cup of coffee.
   Q: Where does the conversation most probably take place?

3. M: Could you deliver the fruits by train, please?
   W: Sorry, Sir. For fruits, we only deliver them by air.
   Q: What kind of goods does the man want to deliver?

4. M: My car broke down again, and it is in the repair shop now.
   W: If I were you, I would buy a new one.
   Q: What is the woman's advice?

5. W: Good evening, Beijing Restaurant. May I help you?

Listening Scripts

Yang Kang: must follow the safety instructions at all times.
Yang Kang: I see.
Sally: Then identify all of the possible electric sources that could cause hazards. Never assume that the equipment or system is de-energized. Remember to test it before you touch it. Lock out/tag out machinery or other equipment you will work on. Turn off power before servicing.
Yang Kang: That sounds very useful.
Sally: It is also important to choose the right personal protective equipment (PPE).
Yang Kang: I got it. Thank you very much.

Cheer up Your Ears

Listen and write down what you've heard. Then read and recite till you can use them fluently.

1. If you can't fix it yourself, call in an <u>electrician</u>, or it could be dangerous.
2. It seems that it's necessary for us to know some <u>electricity</u> knowledge in the workshop.
3. Especially the safety <u>rules</u> of electricity must be grasped.
4. Before you work, make sure you are trained in electrical <u>safety</u>.
5. You must follow the safety <u>instructions</u> at all times.
6. Then identify all of the possible electric sources that could cause <u>hazards</u>.
7. Never assume that the equipment or <u>system</u> is de-energized.
8. <u>Lock out</u>/tag out machinery or other equipment you will work on.
9. <u>Turn off</u> power before servicing.
10. It is also important to choose the right personal <u>protective</u> equipment.

Section Three   Passage: Safety Rules in the Workshop

Attention must be paid to safety in order to ensure a safe working practice in factories. Workers must be aware of the dangers and risks that exist all around them: two out of every three industrial accidents are caused by individual carelessness. In order to avoid or reduce accidents, both protective and precautionary measures must be followed while working.

1. Always listen carefully to the production supervisor and follow instructions. Do not use a machine if the supervisor has not shown you safely.

2. Report any damage to the machines to the supervisor immediately lest they cause an accident.

3. All staff shall wear work clothes and badge. In addition, technicians and maintenance personnel shall wear safety glasses.

4. Always wear a protective apron as it will protect your clothes and hold loose clothing such as ties in place.

5. Must wear protective shoes, which can prevent or minimize foot injury. No slippers, sandals or sneakers allowed in the workshop.

6. Always be patient and never rush in the workshop.

7. Do not run in the workshop, you could "bump" into a machine or another person and cause an accident.

8. Bags should not be brought into the workshop as people can trip over them.

9. Know where the emergency stop buttons are in the workshop.

10. Do not operate a machine if you feel uncomfortable.

Cheer up Your Ears

Listen and write down what you've heard. Then read and recite till you can use them fluently.

1. Attention must be paid to safety in order to ensure a safe working practice in factories.

2. Workers must be aware of the dangers and risks that exist all around them: two out of every three industrial accidents are caused by individual carelessness.

3. Always listen carefully to the production supervisor and follow instructions.

4. Report any damage to the machines to the supervisor immediately lest they cause an accident.

5. In addition, technicians and maintenance personnel shall wear safety glasses.

6. Must wear protective shoes, which can prevent or minimize foot injury.

7. Always be patient and never rush in the workshop.

8. Bags should not be brought into the workshop as people can trip over them.

9. Know where the emergency stop buttons are in the workshop.

10. Do not operate a machine if you feel uncomfortable.

Listening

Task Ⅰ. Listen to 5 short dialogues and choose the best answer.

1. W: What do you usually do in the morning?
   M: We go to the nearby lake and walk around it.
   Q: What does the man usually do in the morning?

2. W: Are you ready to order, Sir?
   M: Yes, I'd like fish and chips and a cup of coffee.
   Q: Where does the conversation most probably take place?

3. M: Could you deliver the fruits by train, please?
   W: Sorry, Sir. For fruits, we only deliver them by air.
   Q: What kind of goods does the man want to deliver?

4. M: My car broke down again, and it is in the repair shop now.
   W: If I were you, I would buy a new one.
   Q: What is the woman's advice?

5. W: Good evening, Beijing Restaurant. May I help you?

M: I'd like to book a table for ten tomorrow evening.
Q: What does the man want to do?

Task Ⅱ. Listen to the passage and fill in the blanks with the missing words or phrases.

I promise you are going to enjoy your stay here in our city. This is a beautiful, quiet city where you can <u>relax</u>, sit by the beach, enjoy great meals and feel safe. You can walk into town and enjoy the fountains or <u>walk</u> along the waterside, Please do not swim here. This is not a safe place to swim for its <u>strong</u> undercurrents. Sanya is the place to go if you want to enjoy swimming <u>in the ocean</u>. You can take a short <u>bus</u> from your hotel.

Unit Five Autonomous Cars

Section One Warming up

Listen to the dialogue and find out which car they are talking about.

Sally: Nowadays, automatic driving has become one of the selling points of cars, but in fact, it is not a new concept, and it has a long history of nearly a hundred years.
Yang Kang: Really? Then when did people begin to study autonomous cars?
Sally: In August 1925, a wireless remote control vehicle named American Wonder was officially unveiled. Francis P. Houdina, an electric engineer of the United States Army, controlled the steering wheel, clutch, brake and other components remotely by means of wireless remote control.

Section Two Dialogue: How Far Away Are We from Autonomous Cars?

Listen and role-play the following dialogue.

Sally: Nowadays, automatic driving has become one of the selling points of cars, but in fact, it is not a new concept, and it has a long history of nearly a hundred years.
Yang Kang: Really? Then when did people begin to study autonomous cars?
Sally: In August 1925, a wireless remote control vehicle named American Wonder was officially unveiled. Francis P. Houdina, an electric engineer of the United States Army, controlled the steering wheel, clutch, brake and other components remotely by means of wireless remote control.
Yang Kang: It's hard to imagine. But is it a real autonomous car?
Sally: No. In 1939, General Motors displayed Futurama, the world's first autonomous concept car. This is an electric vehicle guided by radio controlled electromagnetic fields generated by magnetized metal spikes embedded in the road. However, it was not until 1958 that GM realized this concept.
Yang Kang: Then what about Google's self-driving vehicle? Is it very famous?

Sally: Since 2009, Google has been secretly developing driverless car projects. In 2014, Google demonstrated the prototype of driverless cars without steering wheel, accelerator or brake pedal.

Yang Kang: There is no doubt that the future of autonomous vehicles is promising, but in the process of its development, there are also technical and ethical issues. The high-level autonomous driving we expect will take time to develop and mature.

Cheer up Your Ears

Listen and write down what you've heard. Then read and recite till you can use them fluently.

1. Nowadays, autonomous driving has become one of the selling points of cars.

2. It has a long history of nearly a hundred years.

3. In August 1925, a wireless remote control vehicle named "American Wonder" was officially unveiled.

4. Francis P. Houdina is an electric engineer of the United States Army.

5. In 1939, General Motors displayed Futurama, the world's first autonomous concept car.

6. This is an electric vehicle guided by radio controlled electromagnetic fields.

7. Since 2009, Google has been secretly developing driverless car projects.

8. In 2014, Google demonstrated the prototype of driverless cars without steering wheel, accelerator or brake pedal.

9. There is no doubt that the future of autonomous vehicles is promising.

10. The high-level autonomous driving we expect will take time to develop and mature.

Section Three Passage: Google's Self-driving Vehicles

Google's self-driving vehicles understand where they are and what's around them through sensors that are purpose-built to help the vehicles perceive their surroundings accurately, and software that processes the information received.

1. Laser sensor.

This sensor gives the vehicle a 360° understanding of its environment so the vehicle can sense objects in front of, beside, and behind itself at the same time. The laser also helps the vehicle to determine its location in the world.

2. Safety drivers.

Drivers also test the vehicles daily, reporting feedback on how to make the ride more safe and comfortable.

3. Processor.

Information from the sensors is cross-checked and processed by the software so that different objects around the vehicle can be sensed and differentiated accurately, and safe driving decisions can then be made based on all the information received.

Listening Scripts

  4. Position sensor.

  This sensor, located in the wheel hub, detects the rotations made by the wheels of the car to help the vehicle understand its position in the world.

  5. Orientation sensor.

  Similar to the way a person's inner ear gives them a sense of motion and balance, this sensor, located in the interior of the car, works to give the car a clear sense of orientation.

  6. Radar.

  This sensor detects vehicles far ahead and measures their speed so that the car can safely slow down or speeds up with other vehicles on the road.

Cheer up Your Ears

Listen and write down what you've heard. Then read and recite till you can use them fluently.

  1. Laser gives the vehicle a 360° understanding of its environment.

  2. The laser also helps the vehicle to determine its location in the world.

  3. Google's self-driving vehicles understand where they are and what's around them through sensors that are purpose-built to help the vehicles perceive their surroundings accurately, and software that processes the information received.

  4. Drivers also test the vehicles daily, reporting feedback on how to make the ride more safe and comfortable.

  5. Information from the sensors is cross-checked and processed by the software so that different objects around the vehicle can be sensed and differentiated accurately.

  6. Safe driving decisions can then be made based on all the information received.

  7. Position sensor, located in the wheel hub, detects the rotations made by the wheels of the car.

  8. It helps the vehicle understand its position in the world.

  9. Orientation sensor is similar to the way a person's inner ears give them a sense of motion and balance.

  10. This sensor, located in the interior of the car, works to give the car a clear sense of orientation.

Listening

Task Ⅰ. Listen to 5 short dialogues and choose the best answer.

  1. M: Susan, do you know how long it takes to apply for a visa for China?

   W: 5 – 7 work days, I'm afraid.

   Q: What are the two people talking about?

  2. M: May I take your order, madam?

   W: Yes, I like a vegetable soup and Peking Duck, please.

Q: Where does the conversation most probably take place?

3. W: I am calling to ask about the apartment you advertised in yesterday's newspaper.

M: Yes, it's still available. Would you like to take a look?

Q: Where does the woman get the information of the apartment?

4. W: When can you finish your project report, David?

M: At least in two weeks. I have to check all the statistics again.

Q: What does the man have to do with his report?

5. M: Do you think Mr. Brown is qualified for this position?

W: Yes, he would be a perfect choice.

Q: What does the woman mean?

Task Ⅱ. Listen to the passage and fill in the blanks with the missing words or phrases.

Good evening, ladies and gentlemen! On behalf of our company, I'd like to thank you for coming to <u>celebrate</u> the opening of our new branch office in Hattiesburg. This branch is the 10th office we have <u>set up</u> in the country. I'm glad we finally opened a branch in the southeast area. Now, I would like to <u>take the opportunity</u> to thank all the staff here for your efforts to establish the branch. In order to successfully operate the branch, we need the <u>support</u> of customers like you being present. We will do our best to provide you with the <u>best service</u>. Thank you very much.

Unit Six High Speed Trains

Section One Warming up

Listen to the dialogue and find out when the first high speed train appeared in China.

Yang Kang: When and where did the first high speed trains appear?

Sally: The first high speed trains appeared as early as 1933 in Europe and the U.S.

Yang Kang: Is Shinkansen a kind of high speed trains?

Sally: Yes, in the mid-1960s, Japan introduced the world's first high volume high speed train that operated with a standard (4 ft) gauge. It was called the Shinkansen and officially opened in 1964.

Yang Kang: Then, do you know when the high speed trains first appeared in China?

Sally: In 2005, Beijing-Tianjin Intercity Railway.

Section Two Dialogue: History of High Speed Trains

Listen and role-play the following dialogue.

Yang Kang: What are you doing, Sally?

Sally: I am reading a book about high speed trains.

Yang Kang: What kind of locomotives can be called the high speed trains?

Sally: Well, there are different standards of what constitutes high speed trains based on the train's speed and technology used however. In the European Union, high speed trains are those which travels 125 mi/h or faster, while in the United States it is those that travel 90 mi/h or faster.

Yang Kang: When and where did the first high speed trains appear?

Sally: The first high speed trains appeared as early as 1933 in Europe and the U. S.

Yang Kang: Is Shinkansen a kind of high speed trains?

Sally: Yes, in the mid-1960s, Japan introduced the world's first high volume high speed train that operated with a standard (4 ft) gauge. It was called Shinkansen and officially opened in 1964. It provided rail service between Tokyo and Osaka at speeds of around 135 mi/h. These are the early development of the high speed trains.

Yang Kang: Then, do you know when the high speed trains first appeared in China?

Sally: In 2005, Beijing-Tianjin Intercity Railway.

Cheer up Your Ears

Listen and write down what you've heard. Then read and recite till you can use them fluently.

1. Then do you know when the high speed trains first <u>appeared</u> in China?

2. It <u>provided</u> rail service between Tokyo and Osaka at speeds of around 135 mi/h.

3. These are the early <u>development</u> of the high speed trains.

4. Japan <u>introduced</u> the world's first high volume high speed train that operated with a standard (4 ft) gauge.

5. While in the United States it is those that travel 90 mi/h or <u>faster</u>.

6. In the European Union, high speed trains are that which <u>travel</u> 125 mi/h or faster.

7. There are different standards of what constitutes high speed trains based on the train's speed and <u>technology</u> used.

8. What kind of locomotives can be <u>called</u> the high speed trains?

9. I am reading a book about <u>high speed trains</u>.

10. Beijing-Tianjin <u>Intercity</u> Railway was the first high speed trains that appeared in China?

Section Three Passage: High Speed Trains in China

It was once Japan that was famous for its high speed train network, introducing the world to its Shinkansen, or bullet trains, way back in 1964. But while Japan is still a leader in rail technology, it is now China that holds the crown of high speed train capital of the world.

In twenty years or so since China put into operation its first high speed passenger trains, the country has constructed more than 40,000 km of high speed rail track to create the longest network on Earth. In 2017, the country launched the world's fastest high speed train, known as Fuxing, which travels at up to 350 km/h, reducing travel time between Beijing and Shanghai to four and a half hours.

Now, China's high speed trains—officially defined as passenger trains that travel at speeds of 250 – 350 km/h—take travelers to almost all of the country's provinces. With Inner Mongolia's first high speed line opening in July 2017, only Xizang and Ningxia currently lack high speed trains. But with plans for the continued expansion of the network it won't be long until they too are serviced by high speed lines.

All this makes for a super-fast, and relatively inexpensive way, to cover this country's vast distances.

Cheer up Your Ears

Listen and write down what you've heard. Then read and recite till you can use them fluently.

1. All this makes for a super-fast, and relatively inexpensive way, to cover this country's vast distances.

2. With plans for the continued expansion of the network it won't be long until they too are serviced by high speed lines.

3. With Inner Mongolia's first high speed line opening in July 2017, only Xizang and Ningxia currently lack high speed trains.

4. Now, China's high speed trains—officially defined as passenger trains that travel at speeds of 250 – 350 km/h.

5. China's high speed trains take travelers to almost all of the country's provinces.

6. Fuxing travels at up to 350 km/h, reducing travel time between Beijing and Shanghai to four and a half hours.

7. In 2017, China launched the world's fastest high speed train, known as Fuxing.

8. In twenty years or so since China put into operation its first high speed passenger trains.

9. The country has constructed more than 40,000 km of high speed rail track to create the longest network on Earth.

10. But while Japan is still a leader in rail technology, it is now China that holds the crown of high speed train capital of the world.

Listening

Task I. Listen to 5 short dialogues and choose the best answer.

1. W: Youth Travel Agency. What can I do for you, Sir?

   M: I've read your advertisement about the tour to China; I'm calling to inquire about it.

   Q: What does the man want to know?

2. M: I bought this washing machine last week, but it isn't working now.

   W: Have you brought the receipt with you?

Q: What does the woman want to see?

3. M: Oh, I have left my smart phone in the office.

   W: Don't worry. Let's go back for it.

   Q: What will the two speakers probably do?

4. M: Marry, would you please work out a schedule for tomorrow's meeting?

   W: All right. I'll do it right away.

   Q: What will the woman probably do?

5. M: How was your winter vacation, Amy?

   W: Great, I spent the whole time working as a volunteer in the museum.

   Q: What do we know about the woman during the vacation?

Task Ⅱ. Listen to the passage and fill in the blanks with the missing words or phrases.

ABC Travel Agency organized a 10-day tour for us to many famous places of interest in China in October last year. They arranged for internal travel by high speed train, booked hotels and various guided activities. But we arranged our own flights to and from China and extensions to the tour to Singapore. ABC Travel Agency was good value for money when compared with other travel agencies. It was about 40% less than I was quoted by well-known UK travel companies for the same itinerary. I would have no hesitation recommending it. Its guides were friendly and generally knowledgeable. Most of them spoke good English. Some even went beyond the agreed itinerary and arranged additional activities for us.

Unit Seven Robots

Section One Warming up

Listen to the following material and find out which type of robot is mentioned.

Vacuuming can be a never-ending task, especially if you have pets.

Robotic vacuums have been making the chore easier for more than a decade now.

The pros of robot vacuum cleaner are as follows:

Scheduled cleaning;

Clean in the places where humans can't easily reach;

Spot cleaning;

Less noise;

Plugging itself into its charging station when running out of battery;

Identifying obstacles.

Section Two Dialogue: Do You Like Robots to Work at Your Home?

Listen and role-play the following dialogue.

Yang Kang: Congratulations on your getting the job, Sally.

Sally: Thank you. But I am quite exhausted every day. I am eager to enjoy a more relaxing life after work.

Yang Kang: Then, you need a household service robot. I am obsessed with robots. Don't you think it is a magic that they can fully understand the instructions we give to them?

Sally: Do you like robots to work at your home?

Yang Kang: Sure. I am pretty content with robots working at my home doing things that I really don't like. I can set up the vacuum cleaner to do the chore. Multi-function robotic cookers are able to fry, steam, bake, slow cook, and perform any other action.

Sally: Wow, it sounds awesome. Loud-connected home robots are already becoming a part of your life. They are super helpers and make life easier.

Yang Kang: It is essential that the robots keep your home safe and secure. Smart devices can connect with each other. A robot can be the control center for everything from your coffee machine to locking doors. And you can access it from anywhere in the world!

Sally: Yes, I can control the automatic pet feeder with my cell phone. A household service robot is the perfect companion letting me keep an eye on the cats' condition without having to worry about them.

Cheer up Your Ears

Listen and write down what you've heard. Then read and recite till you can use them fluently.

1. He is eager to enjoy a more <u>relaxing</u> life after work.

2. He is <u>obsessed</u> with robots because they can fully understand the instructions people give to them.

3. He is <u>content</u> with robots doing the housework.

4. You can set up the vacuum cleaner to do the <u>chore</u>.

5. <u>Multi-function</u> robotic cookers are able to fry, steam, bake and slow cook.

6. <u>Loud-connected</u> home robots are super helpers and make life easier.

7. It is <u>essential</u> that the robot can keep your home safe and secure.

8. A robot can be the control center for everything and you can <u>access</u> it from anywhere.

9. You can control the <u>automatic</u> pet feeder with a cell phone.

10. A <u>household</u> service robot can help people keep an eye on the cats' condition.

Section Three Passage: Industrial Robots

An industrial robot is an automatically controlled, reprogrammable, multipurpose manipulator.

Listening Scripts

The most common use of industrial robots is for simple and repetitive industrial tasks, including assembly line processes, picking and packing, welding, and similar functions. They offer reliability, accuracy, and speed. Workers are often exposed to harsh or difficult conditions such as heights, excess dust, loud noise or toxic gases. By introducing robots, efficiency can be improved and the safety of workers is guarded. In the case of detection of defects in industrial products, traditional manual inspection is expensive and inspectors may miss problems or make mistakes, while using machine vision to detect defects can greatly improve efficiency and accuracy. Also, industrial robots have big advantages over people in terms of standardization, safety and intelligence.

As development progresses, robots are expected to replace even more human work and, in effect, help replenish the labor force. Robotics is particularly important to promote the transformation from traditional manufacturing to smart manufacturing. Robots have become the core equipment to make smart manufacturing happen and an integral part of smart factories. The robotics industry is an important symbol of a country's technological strength and level of high-end manufacturing.

Cheer up Your Ears

Listen and write down what you've heard. Then read and recite till you can use them fluently.

1. The most common use of industrial robots is for simple and <u>repetitive</u> industrial tasks.
2. Industrial robots can be used in assembly line <u>processes</u>, picking and packing and welding.
3. They offer reliability, <u>accuracy</u>, and speed.
4. Workers are often <u>exposed</u> to harsh or difficult conditions.
5. <u>Efficiency</u> can be improved and the safety of workers is guarded by introducing robots.
6. Industrial robots have big advantages over people in terms of <u>standardization</u>, safety and intelligence.
7. Robots will replace even more human work and, in effect, help replenish the <u>labor force</u>.
8. Robotics is particularly important to promote the <u>transformation</u> from traditional manufacturing to smart manufacturing.
9. Robots have become the core equipment to make smart <u>manufacturing</u> happen and an integral part of smart factories.
10. The robotics industry is an important symbol of a country's <u>technological</u> strength and level of high-end manufacturing.

Listening

Task Ⅰ. Listen to 5 short dialogues and choose the best answer.

1. M: Excuse me, may I use this printer?
   W: Sorry, it's out of order.
   Q: What does the woman mean?

2. M: Hi, Mary. I've got a new job. The salary is good.

W: Really? Congratulations.

Q: Why does the woman congratulate the man?

3. W: Can I help you, sir?

   M: Yes, I want to rent a car for one week.

   Q: What does the man want to do?

4. W: Good morning. What's the problem?

   M: I'd like to change this shirt for a larger size.

   Q: Why does the man want to change the shirt?

5. W: Hello, Sales Department.

   M: Hello, I'm John Smith from ABC Company. May I speak to your manager?

   Q: Whom does the man want to speak to?

Task Ⅱ. Listen to the passage and fill in the blanks with the missing words or phrases.

I think we'll begin now. First I'd like to welcome you all and thank you for your coming, especially at such short notice. I know you are all very busy and it's difficult to take time away from your daily tasks for meetings. As you can see on the agenda, today we will focus on the upcoming tourist season. First we'll discuss the groups that will be coming in from Germany. After that, we'll discuss the North American Tours, followed by the Asian tours. If time permits, we will also discuss the Australian tours which are booked for early September. Finally, I'm going to request some feedback from all of you concerning last year's tours and where you think we can improve.

Unit Eight A Smart City

Section One Warming up

Listen to the dialogue and find out which city above they are talking about.

Sally: Many cities have made positive contributions to the construction of smart cities and the upgrading of urban intelligence.

Yang Kang: Which city is the most famous one?

Sally: In Helsinki, you can use a smart phone to input your current location and destination. The application can intelligently arrange your journey and provide you with different options of route, travel time and price.

Yang Kang: Really, it's amazing.

Sally: Yes, but there's something more interesting. Residents install sensors in the refrigerator at home to remind the expiration date of food through applications, and advise residents to use food more properly instead of throwing it away after the expiration date.

Listening Scripts

Section Two　Dialogue: What Does a Smart City Look Like?

Listen and role-play the following dialogue.

(Yang Kang and Sally are at the China International Smart City Expo.)

Sally: Would you like to live in a city where buildings turn the lights off for you, there self-driving cars find the nearest parking space, and where even the rubbish bins know when they're full?

Yang Kang: Of course. But what exactly does a smart city look like?

Sally: A smart city refers to a metropolitan or cosmopolitan city that utilizes internet of things technology to effectively manage thecity's assets and resources.

Yang Kang: What are the characteristics that define a smart city?

Sally: Each smart city will have their own unique objective, but they all acknowledge the importance of IOT technology and understand that emerging IOT and cloud capabilities offer a meaningful opportunity to better understand the intimate workings of an urban center.

Yang Kang: How can cities become smart?

Sally: The key for a city to become smart lies in how it can leverage both IOT technology and location-awareness.

Yang Kang: Can any city become smart?

Sally: Any city can become a smart city, regardless of size, as long as it lays the key supporting infrastructure and has the commitment of citizens, municipal officials and local government and collaboration between public and private companies with the different levels of government.

Cheer up Your Ears

Listen and write down what you've heard. Then read and recite till you can use them fluently.

1. Would you like to live in a city where buildings turn the lights off for you?

2. There self-driving cars find the nearest parking space.

3. This is the city where even the rubbish bins know when they're full.

4. But what exactly does a smart city look like?

5. A smart city refers to a metropolitan or cosmopolitan city that utilizes internet of things technology to effectively manage the city's assets and resources.

6. What are the characteristics that define a smart city?

7. Each smart city will have their own unique objective.

8. The key for a city to become smart lies in how it can leverage both IOT technology and location-awareness.

9. Any city can become a smart city, regardless of size.

10. Emerging IOT and cloud capabilities offer a meaningful opportunity to better understand the intimate workings of an urban center.

Section Three   Passage: Smart Cities and Normal Cities: What Are the Differences?

Smart cities are a hot buzzword right now as cities all over the world look to how they can use new technologies. These cities are no longer futuristic scenarios dreamed up by creative thinkers. Instead, real places around the globe are discovering innovative ways to incorporate smart technology into people's everyday lives. Let's take a close look at a smart city and normal city. What are the differences between them? In other words, what makes a smart city smart?

The conceptualization of smart city may vary from city to city and country to country depending on the level of development. But simply put, a smart city is an urban area that uses information and communication technologies (ICT) to ease up the livelihood of its people. It is a municipality that uses ICT to augment operational efficiency, share information with the public and improve both the quality of government services and citizen welfare.

The main purpose of smart city is to create a society which can perform effectively and efficiently making effective use of city infrastructures through artificial intelligence. It also focuses to optimize city functions and drive economic growth while improving quality of life for its citizens using smart technology and data analysis.

Cheer up Your Ears

Listen and write down what you've heard. Then read and recite till you can use them fluently.

1. Smart cities are a hot buzzword right now all over the world.

2. These cities are no longer futuristic scenarios dreamed up by creative thinkers.

3. Instead, real places around the globe are discovering innovative ways to incorporate smart technology into people's everyday lives.

4. The conceptualization of Smart City may vary from city to city and country to country depending on the level of development.

5. A smart city is an urban area that uses Information and Communication Technologies to ease up the livelihood of its people.

6. A smart city share information with the public and improve both the quality of government services and citizen welfare.

7. The main purpose of Smart City is to create a society which can perform effectively and efficiently making effective use of city infrastructures through artificial intelligence.

8. It also focuses to optimize city functions and drive economic growth while improving quality of life for its citizens using smart technology and data analysis.

9. Let's take a close look at a smart city vs. a normal city.

10. It is a municipality that uses ICT to augment operational efficiency.

Listening

Task Ⅰ. Listen to 5 short dialogues and choose the best answer.

1. W: George, do you have any idea to improve our brand image?
   M: Oh, I've just talked about the brand image with Susan.
   Q: What are the two speakers talking about?

2. M: Hello, Customer Service. How can I help you?
   W: Yes, I'd like some information on your telephone banking service.
   Q: What is the woman asking about?

3. M: Excuse me, how can I get to the manager's office?
   W: Take the lift to the fifth floor. It's the third office on the left.
   Q: Where is the manager's office?

4. M: Would you help me to move into my new house this Saturday?
   W: I am really sorry. I have an appointment with my doctor.
   Q: What does the woman mean?

5. W: Could you go to the airport to meet Mr. Smith, the new engineer?
   M: Certainly. What's the flight number?
   Q: What does the woman ask the man to do?

Task Ⅱ. Listen to the passage and fill in the blanks with the missing words or phrases.

Good afternoon passengers. This is the pre-boarding announcement for flight 89B to Moscow. We are now <u>inviting</u> those passengers with small children, and any passengers requiring special assistance, to begin boarding at this time. Please have your <u>boarding pass</u> and identification ready. Regular boarding will begin in approximately ten minutes time. Thank you.

This is the final boarding call for passengers Eric and Fred Collins booked on flight 89B to Moscow. Please proceed to <u>gate</u> immediately. The final checks are being <u>completed</u> and the captain will order for the doors of the aircraft to close in approximately five minutes time. I <u>repeat</u>. This is the final boarding call for Eric and Fred Collins. Thank you.